EXPLORING OUR WORLD

EXPLORING OUR WORLD

CONSULTANT EDITOR: TONY HARE

FOREWORD BY THOMAS E. LOVEJOY

Exploring our World

First published in the U.S. as *Habitats* in 1994, by Macmillan, a Prentice Hall Macmillan Company, New York.

This edition published in 1999 for
SMITHMARK Publishers
A division of U.S. Media Holdings, Inc.
115 West 18th Street
New York, NY 10011

SMITHMARK books are available for bulk purchase for sales promotion and premium use. For details write or call the manager of special sales, SMITHMARK Publishers, 115 West 18th Street, New York, NY 10011.

Created and produced by
Duncan Baird Publishers Ltd
Sixth Floor, Castle House
75–76 Wells Street
London W1P 3RE

ISBN: 0-7651-1027-X

For copyright of photographs, see page 142

10 9 8 7 6 5 4 3 2 1

Library of Congress Catalog Card Number: 98-75010

Printed in China by Imago Ltd

Typeset in Times NR MT
Panorama Computerization: Blackjacks, London; Colourscan, Singapore
Color origination: Colourscan, Singapore

Senior Editor: Clifford Bishop
Design: Steve Painter
Photo Composite Design: Tony Cobb
Picture Research: Jan Croot
Artwork: Ed Stewart, Ron Haywood
Type formatted by Sheena Leng
Component Transparencies for Panoramas supplied by the Frank Lane Picture Agency, the Natural History Photographic Agency, and Survival Anglia Ltd

Contents

Foreword by Thomas E. Lovejoy 6
Smithsonian Institution, Washington, D.C.

Introduction

Types of Ecosystem 8

The Distribution of Species 10

Natural Selection 12

The Ecosystem Budget 14

Competition and Mutualism 16

Succession 18

Islands 20
Dr Peter Moore, King's College, London (pp.8–21)

The Urban Ecosystem 22
Dr Oliver Gilbert, University of Sheffield, England

Habitats in Focus 24

Taiga 26
Jonathan Elphick, Natural History Author

Tundra and Polar 34
Professor Cedric Milner, University of Wales, Bangor

Oceans 42
Dr John Pernetta, Netherlands Institute for Sea Research

Coastlines 50
Dr John Pernetta, Netherlands Institute for Sea Research

Swamps, Bogs, and Mangroves 58
Stephen Mustow, Environmental Scientist

Temperate Grasslands 66
Burkhard Bilger, Earthwatch USA

Temperate Forests 74
Jonathan Spencer, English Nature

Savanna 82
Professor Cedric Milner, University of Wales, Bangor

Rain Forest 90
Professor Bob Johns, Royal Botanical Gardens, Kew, London

Mountains 98
Professor Cedric Milner, University of Wales, Bangor

Rivers and Lakes 106
Dr Mark Everard, National Rivers Authority, England

Deserts 114
Dr Ralph Oxley and Craig Downer, University of Wales, Bangor

Scrublands 122
Nick Turland, Natural History Museum, London

Corals 130
Dr John Pernetta, Netherlands Institute for Sea Research

Glossary and Index 138

Acknowledgments 142

Picture Credits 142

FOREWORD

Back in the Neolithic period, with the dawn of agriculture, our species set in motion a trend which has been the basis of our success but now threatens to bring about our failure. Increasingly, we have perceived ourselves as being independent of the natural world. First, we intensively cultivated a handful of domesticated species, enabling permanent settlements; then came urbanization; then, reinforcing this, came industrialization, and the recent massive shifts of population from rural to urban settings in much of the developing world.

As a compensating gesture, concerns about the loss of habitat led to the creation of the world's first national park, Yellowstone, in 1872; to a list of protected areas which continues to increase; and, more recently, to the International Convention on Biological Diversity and the Framework Convention on Climate Change which were agreed at the Rio de Janeiro summit in 1992. It is now apparent that the old conservation model – of nature being a discrete area with a fence around it while all remaining landscape is free for human activity – will not succeed. Where biological diversity survives in isolated pockets, even natural climate change (which will come as surely as the planets revolve around the sun) can be perilous, as species find themselves unable to disperse in pursuit of required climatic conditions.

To conserve nature, and the complex interrelationships beween species which this book celebrates, a shift is needed to a different model – one in which we acknowledge that we live within ecosystems, and that our activities can be carried out in ways which maintain ecosystem integrity. What does this mean? It means maintaining ecosystem processes, the natural recycling of elements, energy and water which our economic system treats as if they are free. It means maintaining the flow of goods (products of natural origin, including many medicines), as well as the services (such as the cycling of nutrients and water) and the characteristic biodiversity of the ecosystem in question. This is the emerging concept of ecosystem management. As long as the unit of landscape or seascape is relatively large-scale, and as long as the approach is invoked early rather than late in the process of human influence, there is scope and flexibility for our species to live comfortably and constructively with nature, rather than in an atmosphere of polarization and conflict.

Another critically important new approach is that of "adaptive management". Good management, of course, has always been adaptive. What is novel and so very sensible is the notion of designing management plans as scientific experiments, so that their success or failure can be tested properly. This is not entirely new. UNESCO's Man and Biosphere Reserve System certainly fits the definition: a core area of natural habitat provides the standard against which manipulation of the surrounding ecosystem can be compared. Adaptive management may seem unremarkable, but the reality is that the history of our relationship to nature has largely been one of seat-of-the-pants experimentation with little ability to judge results. We can no longer treat the natural world as if it were a great sandbox in which to play whimsically and at will.

Exploring Our World provides a wonderful primer for understanding the natural landscapes and seascapes as well as the human influences into which these new approaches perforce must fit. Fourteen major ecosystem types are presented comprehensively, yet with vivid detail and clarity. Fundamental processes are illuminated, as well as critical quirks of natural history, such as the key role ingestion of seeds by antelopes can play in plant germination in African savanna.

Inevitably the reader will encounter the intrinsic fascination of the natural world, for the constructs and workings of nature inspire wonder in anyone willing to give them the time of day. This ability of nature to awaken excitement is one of the main reasons for hope that ecosystem management and sustainable development can be accomplished in the face of already awesome human numbers and environmental impacts of demonstrably global scale.

Thomas E. Lovejoy
Assistant Secretary for Environmental and External Affairs
Smithsonian Institution

TYPES OF ECOSYSTEM

Ecosystems work on all kinds of levels. An ocean can be regarded as an ecosystem. So can a tidal rock pool, or even a drop of water. At the other end of the scale, there is the Earth in its entirety – including the atmosphere, the solid globe, and all the plants and animals that live on it. What all these things have in common is that they can be thought of as integrated ecological units produced by the interaction of plants and animals with their environment.

On a global scale, the ecosystem concept is usually, and most usefully, applied to biomes. A biome is essentially a major community of organisms extending over large areas with similar climatic conditions. Deserts are biomes. So, too, are tropical rain forests, tundra, and various others. Each has its own distinctive features – its own architecture of vegetation, its own composition of animal life, and its own contribution to the total global economy of life.

Some bacteria can tolerate extreme conditions, and survive in the most unlikely places – in polar ice, thermal springs, salt lakes, deep-sea thermal vents, solid sedimentary rocks, and even the upper atmosphere. So all these places are part of the *biosphere*, the Earth's realm of life. Most plants and animals are more exacting and cannot survive when the temperature drops permanently below 14°F or rises above 113°F. Indeed, each kind of organism thrives under a very particular set of conditions, so different parts of the globe support different kinds of plants and animals.

However, the great global ecosystems, *biomes*, are characterized by the structure of their vegetation and the forms and functions of the animals that occupy them, since the same biome may contain different species in different parts of the world. Tropical rain forests in the New World contain a collection of species very different from those in the Old World. Yet even though the species may be different, similar plant life-forms can be identified. The dominant form in the rain forest is the tree, whether the rain forest is in Africa or America; the dominant life-form in tundra, wherever it occurs, is the dwarf shrub or cushion-forming perennial plant.

Danish botanist Christen Raunkiaer proposed a system of classifying vegetation according to its life-form, based upon the height above ground (or below ground) of its "perennating" organs – that is, the growing parts, typically the buds. The importance of the perennating

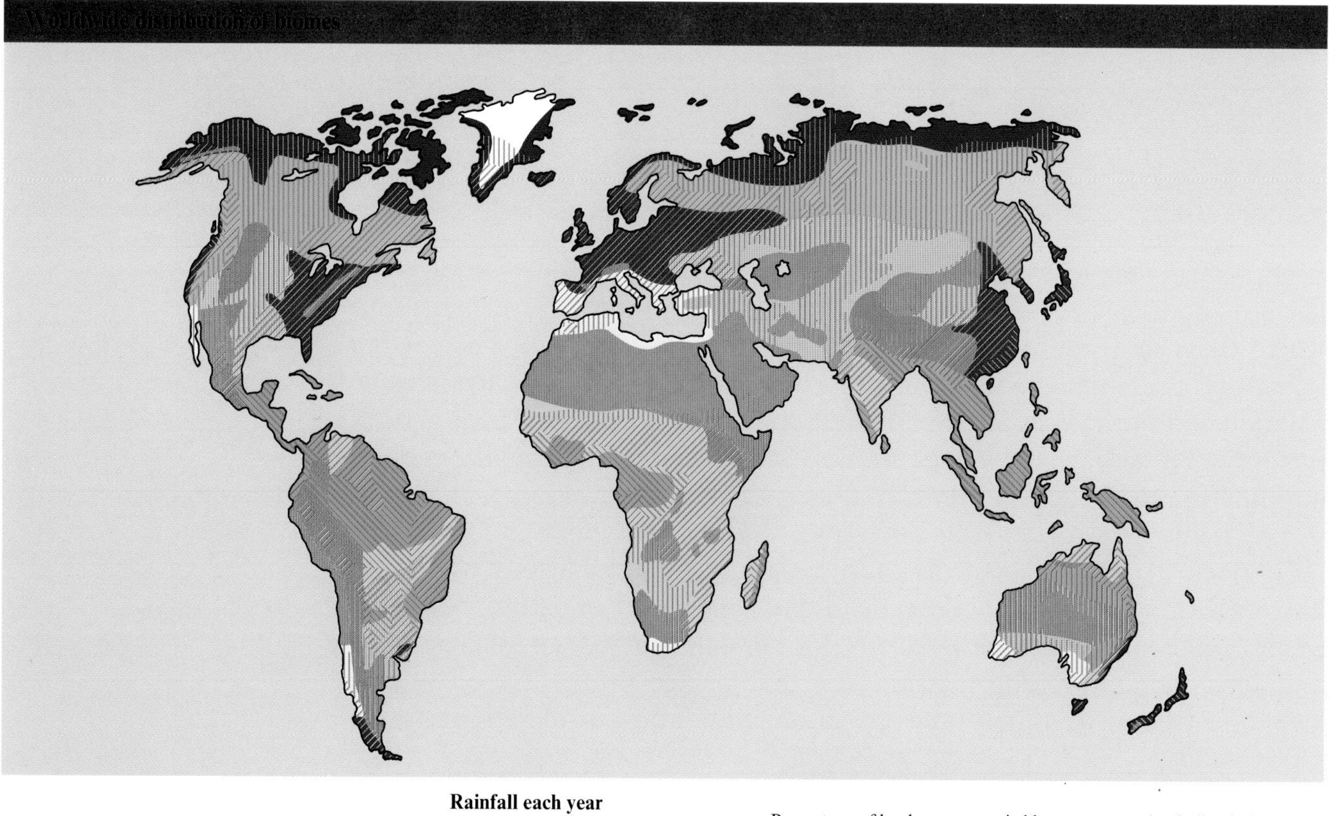

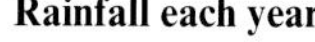

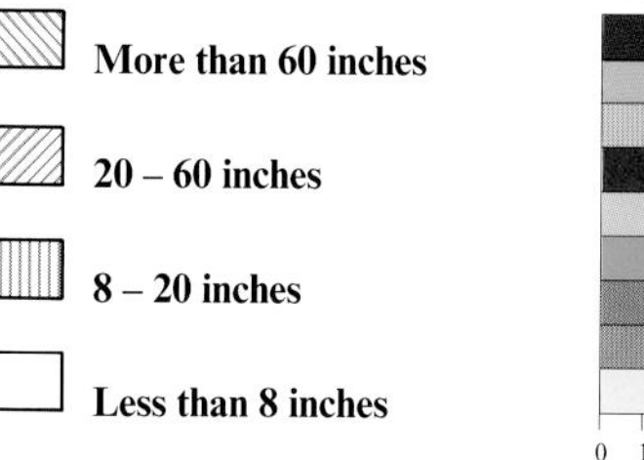

Percentage of land mass occupied by ecosystems (excluding ice)

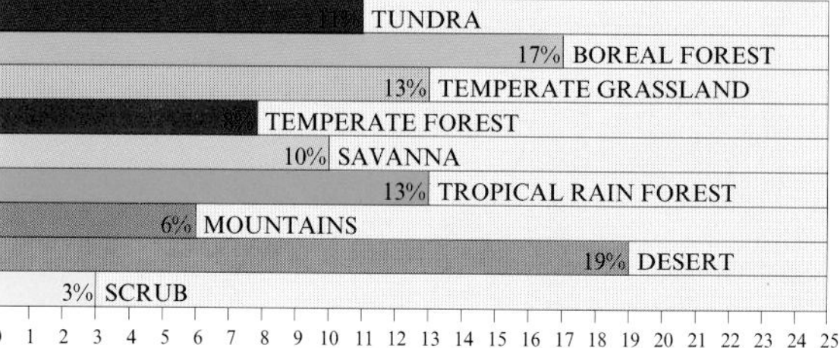

Biome distribution
Each global ecosystem has its own pattern of distribution, determined by a complex pattern of climate. Moisture condenses out of warm, rising air over the equatorial regions, which are thus both hot and wet. The dried air descends again, producing arid belts around the Tropics, and moves poleward meeting cold polar air to give a climatically unstable zone in the temperate regions.

organs is such that they have to be protected to allow survival into the following year. Trees (*phanerophytes*) thus hold their buds high above the ground, and are appropriate life-forms in equatorial regions where moisture is abundant, but are less suited to dry regions, and unsuited to cold regions, being particularly sensitive to frost and wind damage. In the Arctic tundra, the most successful life-form is the plant that does not raise its buds far above the ground – dwarf shrubs and cushion-plants (*chamaephytes*).

In dry regions, plants often survive if they retreat underground and use buried bulbs and tubers to withstand drought (*geophytes*). Or they may adopt an annual lifestyle, enduring an unfavorable period as a seed (*therophytes*). In temperate regions, one of the most successful life-forms is the *hemicryptophyte* ("half-hidden"). Such plants have perennating organs at the surface of the soil, and include temperate herbaceous plants – clover and dandelion.

Each biome has its own set of plant life-forms and is dominated by some of them. Raunkiaer called this a biome's "biological spectrum," and found that it was an effective way of describing the structure of vegetation within the biomes. Another aspect may be the way the vegetation adapts to cope with the vicissitudes of climate. Evergreen leaves can be advantageous in the warmth of the equatorial regions and, in different forms, in the cold of the boreal regions. Between them, the deciduous habit is most effective, enabling the plant to live through seasonal water-shortages. Succulent leaves and stems, or small leathery leaves (*sclerophylls*), are found in arid locations. Life under water for higher plants often demands specialized air-conducting systems to the roots.

Animal life is not usually classified in terms of life-forms, chiefly because it is the nature of each biome to provide particular niches in the environment to which animals may adapt. In a tropical grassland (South America or Africa) there are opportunities for large-bodied grazers, and for large predators to hunt them and scavengers to feed on any leftovers. The same role may have been adopted by different species of animal in the various geographical locations of a biome, but the overall structure and function of the ecosystem remains the same. Hence, areas with similar climates in different parts of the world develop the same characteristic appearance, with a variety of plants and animals that have traveled the same evolutionary path.

THE DISTRIBUTION OF SPECIES

Every species has its own characteristic pattern of distribution, with three distinctive zones. There is a central core zone where the species is firmly established and maintains its population securely. Around this is a peripheral zone where the species survives, but in reduced numbers and at risk in the event of natural catastrophe or climatic shift. And there is an outer zone where the species may occasionally stray, but cannot establish itself on a permanent basis.

Various factors affect the limits of these zones. Each species, for example, has its own unique weaknesses that prevent it extending its range, such as poor powers of dispersal. The natural stresses of the environment may also be a significant factor – climate, soil, water quality, and so on. So, too, may be historical factors – that is, mere accidents of the geological and evolutionary past.

Sometimes a species may simply be driven from an area it is otherwise suited to simply because it cannot compete successfully with another species that makes its living in the same way and, under the prevailing conditions, does it better.

If an organism is absent from a specified area of the globe, it means one of three things: it has never managed to travel there (and so has not had the chance to try to establish itself); it is capable of traveling but is not capable of sustaining a population there; it cannot cope with the environmental conditions or the competition from indigenous organisms.

Some species are prisoners of their own evolutionary history – they have evolved in a particular area from which no escape has been possible. Islands, even large ones like Australia, are rich in endemic species that originated there and spread no farther. Species of this type are described as paleoendemic, meaning that they have been resident since prehistory. Other species may be neoendemics: species that have evolved only very recently and although currently restricted to their place of origin may, in due time, spread farther.

Physical barriers often prevent an animal or plant from extending its range. Oceans represent an obvious barrier, but mountain chains or some other form of unfriendly environment can constitute the same sort of obstacle. This may be why some of California's flora is so distinctively localized.

Fragmented populations

The populations of some species, such as the gorilla, seem to be fragmentary, with various groups divided by barriers that have arisen since their evolution. Magnolias are another example: they grow wild in Southeastern Asia and in Central and North America. Magnolias are very ancient flowering plants and were widespread early in the age of the dinosaurs, when all the world's continents were joined into one supercontinent, Pangea. But when Pangea broke up and the continents drifted apart, the magnolias became extinct over much of their former range.

Similarly, the brush mouse of North America is now found mainly in New Mexico, although there are isolated populations in Texas and in the intervening area. In this case, though, climatic changes in the past 20,000 years probably account for the development of this "disjunct" (fragmented) distribution.

Many plants and animals are prevented from extending their range by climate. There is a huge family of palm tree species, for example, but almost all are localized within the tropics and subtropics. Their constituent cells are sensitive to cold, so they do not survive well in areas subject to prolonged frost. In a similar way, the winter range of the eastern phoebe, a North American

Gorilla distribution
There are two wild gorilla populations: one in the western part of Zaire and the other in the mountains on the Zaire-Uganda border. Undoubtedly these two were once joined, but during the last 2 million years there have been many shifts in climate that have isolated the gorilla's forest habitat. In addition, the Zaire River is thought to have increased in size, so becoming a more formidable barrier. Human activity has also restricted the movement of these shy and reclusive animals.

Zairian gorillas

migratory bird, corresponds to areas south of the 25°F January isotherm. If the January average drops much below this, the bird has to expend so much energy maintaining its own body heat that long-term survival becomes difficult.

Competitive limits

Many species will not spread into an apparently suitable area because they cannot compete with other resident species. Magnolias, for example, can be grown perfectly well in European gardens but do not spread into the wild because they simply cannot compete with the indigenous tree species.

Indigenous species are not always able to resist so forcefully. The introduction of a species from one part of the world to another can cause ecological catastrophe.

The bringing of the rabbit to Australia, for example, was utterly calamitous: it rapidly and successfully took over the limited grassland resources formerly dominated by domestic sheep. There was no animal filling an exactly equivalent ecological niche, which meant that competition was minimal and the rabbit multiplied so rapidly that it became a pest.

The prickly pear and the water fern *Salvinia* proved almost equally problematic as plant pests. In Europe, the American gray squirrel has likewise replaced the native red squirrel throughout most of England since its introduction in the late 19th century.

Human intervention

Humans have influenced distribution patterns in other ways than introductions. Their role in shaping habitats and eliminating unwanted species has also been crucial.

In the steppes of Eurasia, for example, the various species of bustard are threatened not only directly by human hunting activities, but also by the grazing and plowing up of their natural habitat.

The steppes are also home to some valuable plant resources – such as the ancestors of present-day wheat and barley – which are similarly becoming restricted in distribution by habitat change. The loss of such species would rob the world of valuable genetic reserves that might help improve current domestic strains, perhaps by increasing resistance to drought, to soil salinity, or to organic pests.

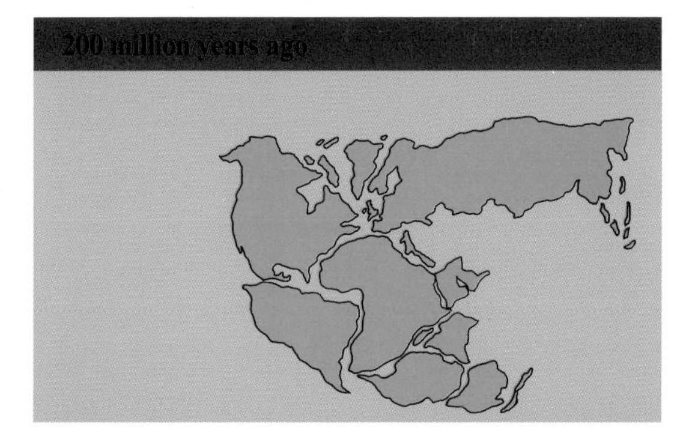

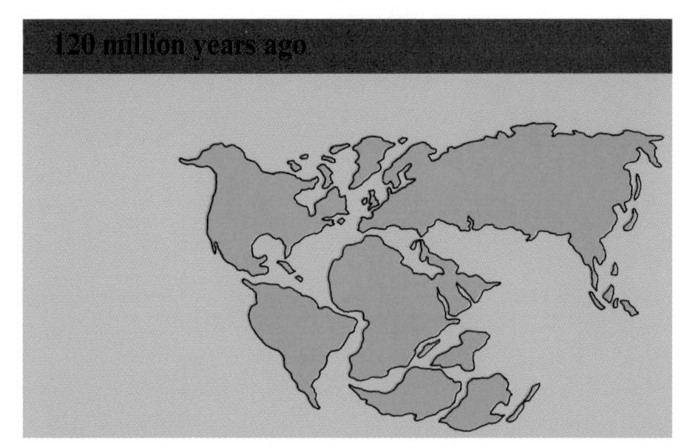

Continental drift
One of the most significant factors underlying present-day distribution patterns of animals and plants is the geological process called continental drift. The map of the Earth has not always looked the way it does today: 400 million years ago there was just one vast supercontinent. The subsequent shifting of the Earth's crust due to geophysical convection currents in the mantle has caused considerable splitting, traveling, and rejoining of these land masses, shown here in the early Jurassic (top left) and early Cretaceous (bottom left) periods.

These movements have influenced the distribution of species around the world and contributed to the way in which the world's flora and fauna can be divided into six major zoogeographic regions (below).

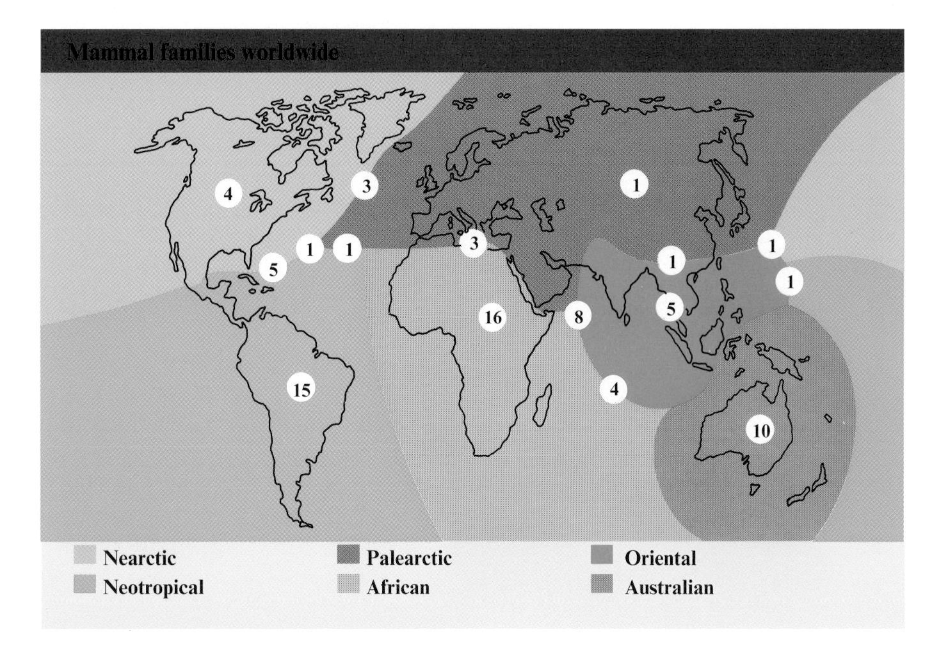

The wandering mammals

Soricids (shrews)
Sciurids (squirrels and chipmunks)
Cricetids (hamsters, voles)
Leporids (rabbits, hares)
Carvids (deer)
Ursids (bears)
Canids (dogs)
Felids (cats)
Mustelids (weasels, badgers)
Bovids (cattle, sheep, antelope)
Murids (rats, mice)

Distribution of mammals
The numbered circles above indicate how many mammal families are restricted to any one zoogeographic region, and how many are distributed over two or more regions. Excluding the 11 "wandering" families that have dispersed all over the world (left), the pattern of distribution reveals that more than half of the families are restricted to one zoogeographic region or another. The early separation of Australia from the rest of the supercontinent is reflected in the fact that all 10 of the Australian mammal families are found nowhere else.

NATURAL SELECTION

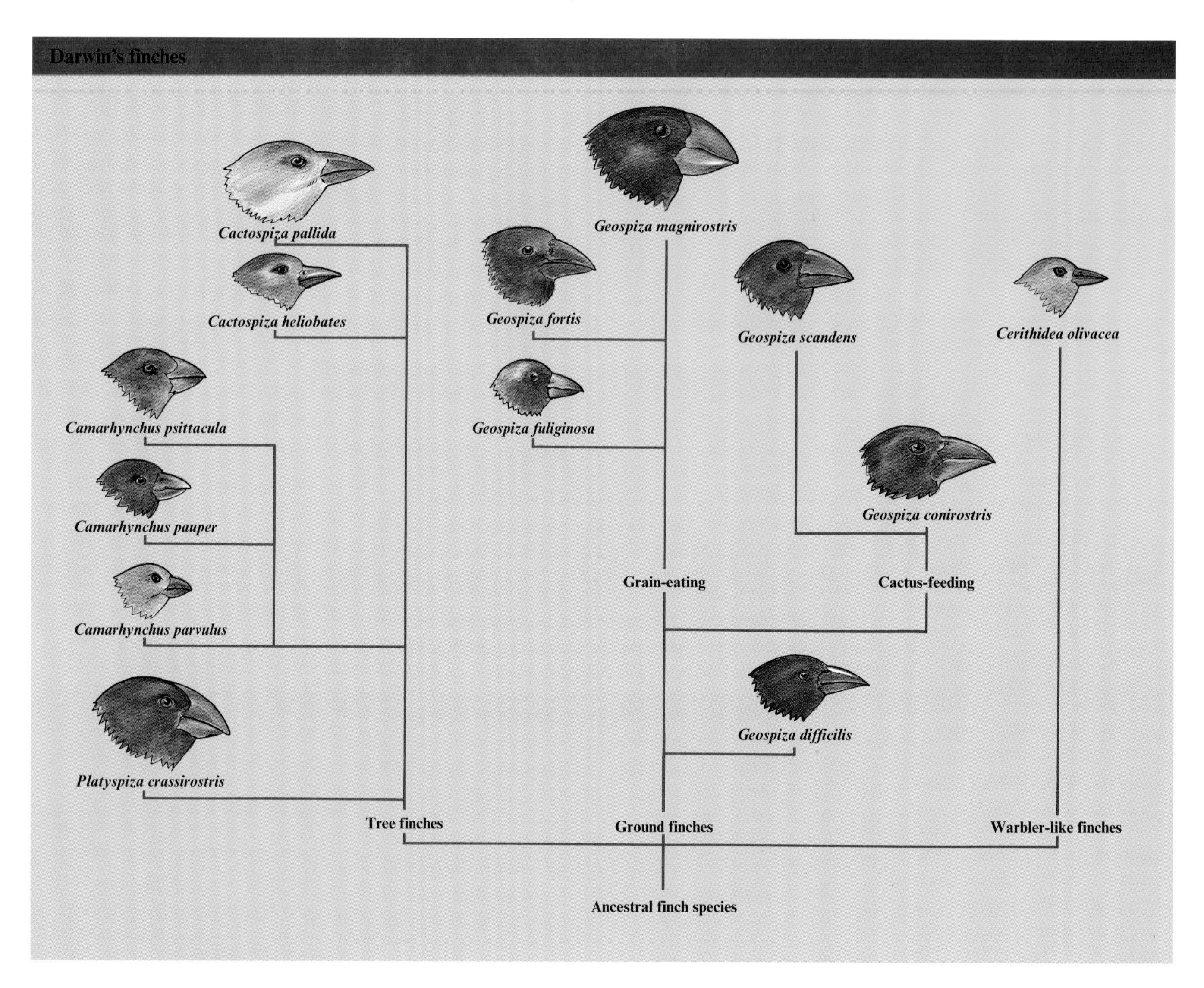

The biological world is continually changing. It is not only the physical environment – the climate, the water level, the shape of the ground, and so on – that varies; the species of animals and plants that occupy the world themselves become modified in time, as a study of the fossil record clearly illustrates.

In fact, the physical and biological worlds are closely related, for a change in physical conditions usually means that the plants and animals in a region are a little less well adapted to their new circumstances, and the process of alteration, or evolution, begins.

The idea that evolution occurs as animals and plants adapt to their environment was already quite familiar when Charles Darwin wrote *The Origin of Species*, but he provided an explanation for how such change actually occurs. The process, which he called natural selection, operates as the stresses of the environment lead to the death of those individuals that are less well adapted and the survival and breeding of those that are more fit to meet the rigors of their world.

Biologists divide the living world into units that they call "species," which are not always easy to identify. In most cases, species are fairly plainly distinguishable from one another in their appearance, behavior and biochemistry. Yet different species may look quite similar. What usually distinguishes a species is that individuals from different species cannot normally interbreed. But even this distinction does not always apply. Some species of oak tree can actually interbreed with other "species" of oak, and horses can interbreed with donkeys, though the product, the mule, is infertile.

The interbreeding of closely related species, but the inability to interbreed of distantly related ones, suggests that "close" species have recently evolved from a common ancestor and the barriers to interbreeding are not yet complete. Evolution often seems to proceed as a result of species splitting, frequently with one of the two products proving more successful than the other. But sometimes they continue to coexist. In Europe, two species of migratory insectivorous birds, the pied and the collared flycatcher, look very similar, overlap in range, rarely interbreed, and one migrates to the west, the other to the east. This behavior pattern may have developed during the last glaciation when one original flycatcher species became separated into two groups by the alpine ice sheets, and

The "vampire finch" *Geospiza difficilis* has earned its name through its habit of feeding on the blood of other birds, such as gulls

The woodpecker finch *Cactospiza pallida* of the Galapagos Islands has learned to use tools such as twigs to probe for insects

the habits they developed have continued to separate the populations.

Change of this sort originates in the variability of populations. It may be difficult to detect subtle differences in the individuals of a population of animals, but they are present. In observing flocks of swans, ornithologists have found that they can distinguish individuals by the patterns on their beaks. There will also be physiological variations such as disease resistance, cold-hardiness, and fertility, which all play a part in an animal's adaptation to its environment. These variations, if they are based in the animal's genes, are the raw material of evolution.

Populations of animals and plants generally produce more young than can be supported by the resources of their environment, resulting in high mortality among the young. Those with slightly different features, making them better equipped to cope with competition for resources, will be more likely to survive, breed, and pass on those favorable features to the next generation. In this way, useful variations are preserved in the population, and the environment (physical and biological) lets through only the most useful of the various attributes. Darwin called this "natural selection." The raw material is generated by chance (genetic changes, recombinations, and mutations), but the selection process determines the direction of evolutionary development.

The result of this process is the increasing specialization of species, which become increasingly equipped to occupy a particular role in the community more efficiently than any competitor species. Different species of wading birds on mudflats in estuaries have different lengths of bills and legs and feed in various depths of water or at different depths in mud. Flamingoes have long legs and can feed in deep water, filtering it with their beaks. Dabbling ducks, such as shelduck, feed in shallow water and "up-end" to feed on the submerged mud surface. Stilts and godwits are able to wade into fairly deep water on their long legs and probe deeply into the mud with their long bills. Redshanks and dunlins have relatively shorter legs and bills and are consequently limited in the depth of water and mud at which they can feed. These adaptations mean that each species exploits parts of the environment at different times and tides, so the competition for food is reduced and species can successfuly coexist.

Darwin's finches

Charles Darwin visited the Galapagos Islands off South America in 1835. He was impressed by 13 species of sparrow-like birds that were all closely related and yet differed in features such as the size and shape of their bills. Some ground-feeding types (species of *Geospiza*) had a range of bill sizes related to the size of seeds they foraged for on the soil surface. Other finches occupied the canopy and fed upon fleshy fruits and insects. The woodpecker finch, *Cactospiza pallida*, impressed Darwin by its use of twigs or cactus spines as tools to extricate insects from crevices.

Darwin eventually accounted for these varied but related finches by suggesting that an original immigrant finch species, arriving in a land where there was little danger from competition used different lines of development permitting the exploitation of all the niches available on the islands.

Recent studies of Darwin's finches have provided clear evidence of populations adapting to new conditions. In the late 1970s there was a prolonged drought on the Galapagos leading to a scarcity of seeds. The population levels of the medium ground finch species, *Geospiza fortis*, declined severely, but the high mortality was not a random affair. The small seeds that were the finches' preferred food disappeared faster than larger seeds. As the finches had to turn to larger, harder seeds, the bigger of the birds with their bigger beaks, found it easier to survive and breed. Continuation of drought could lead to the evolution of a new and larger version of *Geospiza*.

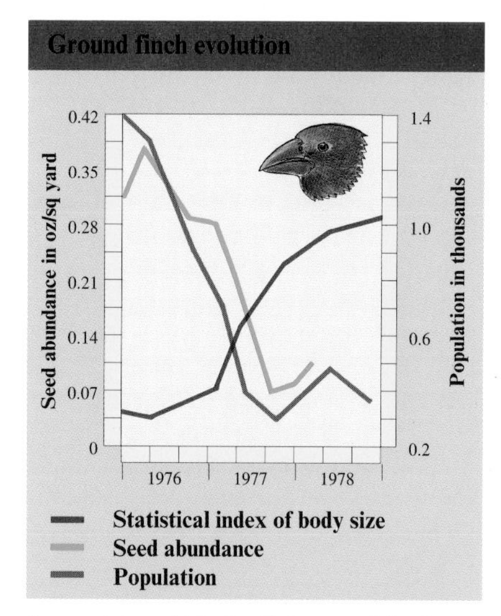

A recent drought in the Galapagos Islands, when rainfall on the 100-acre islet of Daphne Major fell from an average of 5 inches to 1 inch, showed natural selection in action. The population of ground finches declined in line with the increasing scarcity of seeds – but the average size of the finches went up, since bigger finches survived better.

THE ECOSYSTEM BUDGET

All living things are constantly interacting with their non-living environment and with the other plants, animals and microbes that share it. This network of interactions is what we call the ecosystem, and understanding how ecosystems function is essential if we are to be able to manage them properly, to take the food we need out of them, to obtain timber for building and to conserve and protect their wildlife. The idea of the ecosystem can be applied at any scale, from a tussock of grass, to a pond, or to the Sahara Desert.

We need to understand how the whole ecosystem obtains and deals with its energy, how nutrient elements vital for life circulate in the system, and we also need to construct balance sheets of input and output to make sure that more is not taken out of the natural world than is being put into it. Only in this way can the human race achieve a sustainable balance with its environment.

All living organisms need energy to survive. We use energy every time we move, think, even while we sleep. Almost all of this energy comes ultimately from sunlight, but it often reaches us along diverse paths, passing through other organisms on the way. In natural ecosystems these patterns of energy movement can be extremely complicated and closely interlinked as the animals obtain their necessary energy from a range of possible sources. This elaborate series of interactions is called a food web.

The entire structure of the ecosystem is supported by the energy-fixing process of photosynthesis. The ability of green plants to trap the energy of sunlight, which they absorb by means of the pigment chlorophyll, and to store that energy in a stable and useful form by taking the carbon dioxide out of the atmosphere and building it into complex organic molecules, is the means by which almost all ecosystems are supplied with energy. The rare exceptions, such as submarine volcanic vents, are where certain bacteria are able to obtain energy from chemical reactions, like the oxidation of iron compounds, that do not involve sunlight.

Some ecosystems have more efficient energy-gathering capacities than others. They have a higher rate of primary production and hence have the capacity to support more animal and microbial life. Tropical rain forests, coral reefs and wetland swamps have particularly high levels of primary production and are usually rich in species. The energy that is fixed may be consumed by grazing animals (ranging from aphids to cows) or may die and fall to the soil where it is eventually consumed by detritivores, such as springtails and woodlice; or it may be invaded by decomposing fungi and bacteria. All of these organisms use the plant material as a source of energy and ultimately dissipate it as they respire, releasing heat and carbon dioxide back into the atmosphere.

The plant itself needs some energy to absorb elements such as phosphorus, potassium, and nitrogen from the soil and to produce flowers and seeds. It may also store some energy as it grows in size.

Biomass

The energy reservoir within an ecosystem, mainly in the form of vegetation (in terrestrial locations) but also including the bodies of animals and microbes, is termed its biomass. Some high biomass ecosystems, such as the tropical rain forest, are very productive, but this is not always the case. A pool with a rich nutrient supply in a warm climate may be very productive yet contain a relatively low biomass (algae, invertebrates, and fish).

The consumer animals may be consumed by predators, so that energy passes from one level (called a trophic level) to another. Since some energy is lost in respiration at each stage, the total energy available declines as it passes through the food web and this limits the number of steps that are normally found. Chains with more than five steps (such as plant – aphid – beetle – starling – cat) are quite unusual.

A desert food web

In the Great Kavir desert of Iran, woody shrubs, such as *Zygophyllum* and *Artemisia*, together with some grasses, annual herbs and bulb plants such as wild tulips, anenomes and fritillaries, provide the bulk of the very poor primary production. Wild sheep, gazelle, and the increasingly rare wild ass graze directly on this vegetation. Some birds also operate at this trophic level, depending directly on the vegetation for their food; they include bustards, sandgrouse and larks, but some of these, together with many of the small rodents that occupy the desert-shrub areas, prefer the seeds of plants to their leaf tissue, and these are produced in profusion by the shrubs and herbs. All these grazing and foraging animals and birds provide, in turn, a food resource for the large predators of the desert-shrub ecosystem, and each of these has its preferred prey. The big cats take the larger grazers, such as ass, sheep, and bustard. Golden eagles abound and they will also take bustards, together with smaller prey, like sandgrouse and hare. Foxes overlap with them in consuming small mammals, but will also eat invertebrate prey, such as beetles.

A detritus-based food web also exists, many invertebrate animals taking the dead remains of plants and animals, together with the feces of the larger animals, as their source of energy. When the large animals themselves die, either by natural causes or following predation by a large carnivore, the scavenger organisms are never far away. Among the vultures, the griffon often dominates the early stages of feeding on a carcass. When only bones are left, the bearded vulture is able to carry them into the air and drop them so that they break open and expose the marrow. All available energy is thus used to its full extent.

The recycling of resources

Energy is stored within materials, particularly organic materials consisting largely of carbon, hydrogen, and oxygen, but other elements are also necessary to construct living organisms, including nitrogen (used in proteins), phosphorus (in membranes and nucleic acids), potassium, calcium, sulfur, and so on. These elements are obtained by plants mainly from the soil, while animals receive them in their food and drinking water. The decomposing activity of mircrobes releases these elements from dead organic matter into the environment where they become available for re-use. This is in contrast to energy, which is not recycled but is lost as heat. Fortunately, the sun represents a an effectively unlimited supply of external energy.

The soil's reservoir of nutrients, however, may become depleted by plant growth, even allowing for microbial recycling. Some may be lost to the ecosystem as rain water passes through and leaches some elements away. But the rainfall, especially near the sea, may also be a source of new elements, together with the parent rock that fragments and weathers in the soil to release elements that have been locked up sometimes for millions of years. In the case of nitrogen, certain free-living bacteria, together with others that live in symbiotic association with plants, are able to fix it directly from the atmosphere within the soil.

When an ecosystem is exploited for agriculture or forestry, the removal of these elements may take place faster than their natural rate of replenishment, in which case the human manager must make good the deficit with fertilizers. Nitrogen can be fixed chemically but requires considerable investment of energy because high temperatures and pressures are needed. Phosphorus is mined from certain rich deposits, one of which is the fecal droppings (guano) of sea birds, especially from the Pacific coasts of South America. Here humankind is tapping into a natural cycle by exploiting the accumulation of phosphorus by fish-eating birds.

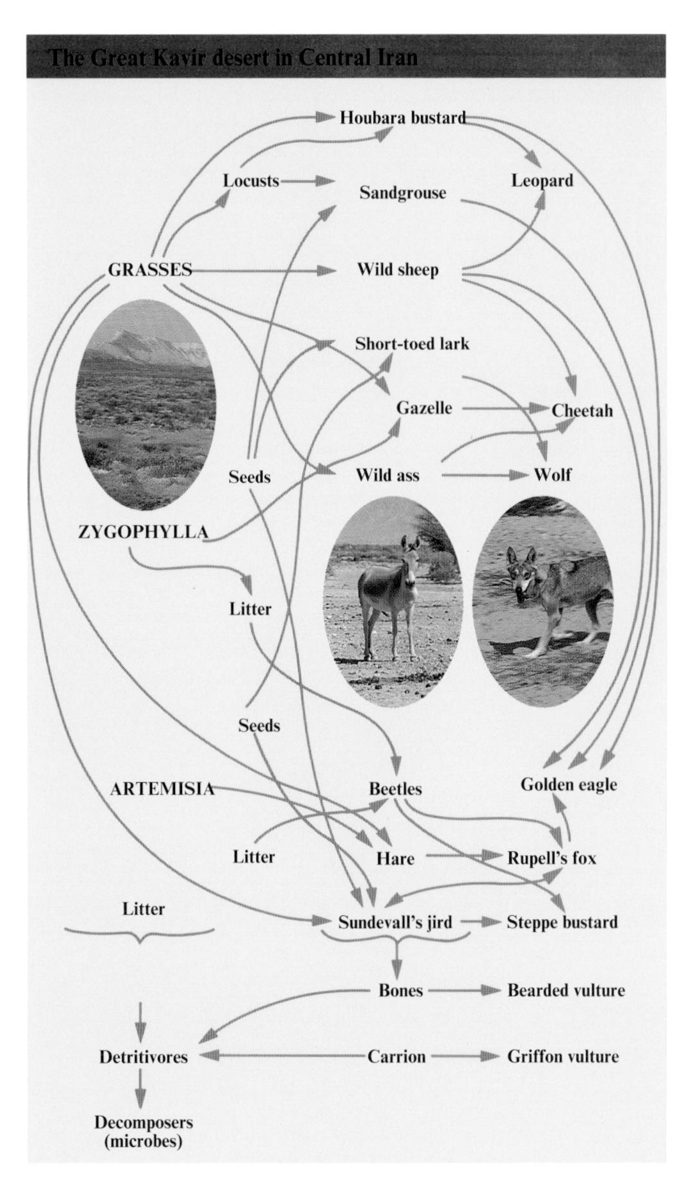

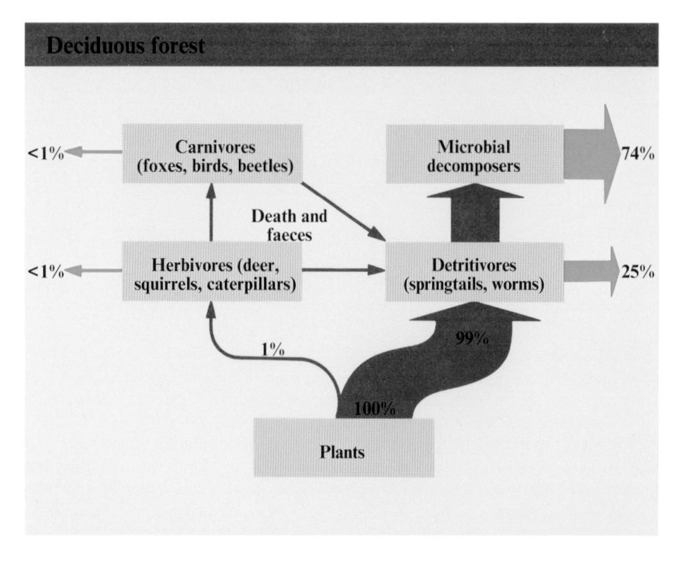

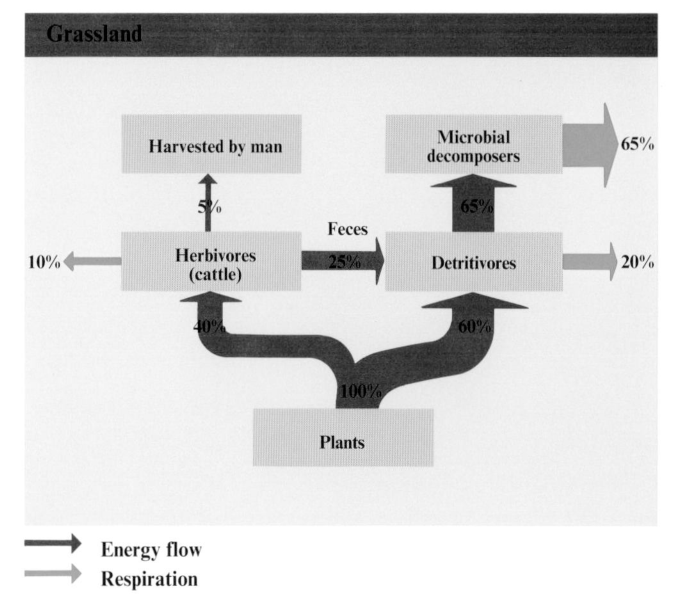

Primary productivity is low in desert environments, such as the Great Kavir desert of Iran, because of poor water availability. However, there are still many species of animals that make good use of the limited energy resources, ultimately exploiting the three primary producers: the grasses, the *Zygophylla,* and the *Artemisia* (as plants, litter, and seeds). The fairly low amount of energy available to the food web seems to have little effect, in this case, on the web's ultimate complexity. By careful quantitative study it is possible to determine just how much energy passes along the various possible channels within an ecosystem.

In temperate grasslands that are grazed by cattle, for example, a significant proportion of the photosynthetic energy fixed by plants (up to about 40 percent) is consumed by the large grazers and part of this (perhaps 5 percent of the total) is harvested by the human manager as either dairy produce or beef. Agricultural ecosystems are very simple compared with almost any other ecosystem. This is because, through the use of pesticides, fences and systematic extermination, humankind has eliminated most of the species that are not economically advantageous. As a result, species diversity is low.

In a forest, by contrast, the grazing section of the food web often represents a very small part of the energy flow, perhaps as little as 1 percent of the total. Instead, the great bulk of the energy passes into the detritivore and decomposer sections of the food web. This energy is effectively lost as far as human interests are concerned, and as a result the human management of forests (apart from their use as timber) has in the past often involved their clearance and the subsequent creation of more easily exploited grasslands.

COMPETITION AND MUTUALISM

In their daily struggle for existence, species have to contend with the stresses occasioned both by their environment and by their interactions with one another. Some, therefore, compete for the limited vegetative resources available, others feed upon each other, and yet others exist in apparent cooperation, involving a system of mutual and reciprocal benefit known as symbiosis or mutualism.

Competition for resources may occur between individuals of the same species (that is, it may be intraspecific) or between different species that have similar requirements (interspecific). Predators kill and consume members of (usually) other species. Parasites, which are generally much smaller than their hosts, are not true predators but nonetheless draw sustenance from their hosts. In symbiosis, there is often a delicate balance between the demands of the two species involved, resulting in benefit to both.

In order to survive, every species must reproduce at a rate sufficient to replace the individuals that die. The birth rate of a stable population is equal to the rate of mortality, whereas in an expanding population the birth rate exceeds the mortality rate. But almost all organisms display a high rate of mortality among the young. The problem is that species overproduce – they give birth to more young than can be sustained by the resources of their environment. Wasteful as this may seem, it ensures that those which do survive are the best adapted to the particular problems that beset that species.

Many factors may restrict the growth of a population: food supply, space, physical requirements such as light or water, and the levels of predation, parasitism, and disease, for example. But the combination of overproduction and of consequent high mortality tends to mean that all individuals in the population of a species must compete with one another. In fact, because all the individuals of a species fundamentally require the same elements from their environment, they generally present a more serious threat to one another than individuals of other species with somewhat different needs.

Such intraspecific competition is evident in a patch of weeds, or in a woodland clearing invaded by birch trees. Large numbers of small individuals gradually give way to smaller numbers of large individuals as the fittest assume dominance in the population. This is called self-thinning.

Competition between species is by no means unknown, of course. Scavengers such as hyenas, jackals and vultures may compete for a carcass, and the order of priority that is then established is decided on the basis of size and degree of aggression. More robust woody plants usually replace herbaceous plants in the course of succession on an abandoned piece of ground.

In this way, interspecific competition can lead, through evolutionary development and adjustment, to a coexistence between species and on to the creation of a community that does not imply cooperation, but exists at an equilibrium maintained by the relative strengths and weaknesses of the different species under different conditions.

Predation pressure

Competition arises when two individuals seek the same resource, but there are other types of interaction. One animal may feed upon another, and so act as a restrictive factor on the population level of its prey. On the whole, predation seems to be less important in determining the maximum population of a species than a restriction on resources (such as food or habitat). However, there are exceptions – especially when the predator is one that has arrived well after the evolution of the prey species (such as predators introduced on to islands).

Predators are usually bigger and more powerful than their prey, and their populations must clearly always remain smaller. If a predator becomes so abundant that the prey population seriously declines, hunting becomes more difficult, and the survival rate among young predators likewise declines.

Parasites are usually much smaller than their hosts, and may live either outside or inside the host's body, deriving sustenance from the host's tissues. Parasites that have evolved in parallel with the host do not cause its death. When a parasite replaces one host with a different species host, however, more serious problems may arise, for the new host may be less well adapted.

From the point of view of a plant, an animal that grazes on it might be regarded as a type of predator. Such an analogy is quite close in the case of seed-feeding animals, which effectively kill members of the plant population when they feed. But grazers do not usually kill

The blood-sucking tick
The tick attaches itself to the outside of the host's body and feeds on blood by means of specialized sucking mouthparts. Such a parasite is not normally fatal to the host, as that would not be in the parasite's interests.

the plant – they only remove a portion of it. In some respects, then, they behave more like parasites than predators.

Many species not only tolerate one another but work to each other's mutual benefit. A few forms of bacteria, for example, normally occupy the human intestines and are essential to the working of the digestive system.

The yucca moth lives only on yucca trees in the deserts of North America. It feeds on the tree's fruits, but in the process of egg-laying it pollinates the flowers, and in this way is essential to the completion of the yucca life-cycle.

But relationships in such symbiotic associations are not always entirely co-operative. The numbers or assortment of human intestinal bacteria can easily be distorted, causing an alimentary upset. Even the yucca moth is subject to rigid population control by the tree, which drops many of the infected fruits so that the moth can never become too numerous. The normal balance is maintained not by mutual altruism but by offensive action on the part of one partner or the other.

Dividing the spoils
Competition between individuals of the same species can lead to problems, especially at times when they live in really close association with each other, such as when breeding. In such a situation, competition may be reduced by what is called resource partitioning – an arrangement by which males and females make slightly different demands on their environmental resources, perhaps consuming different types of food or feeding in different localities. This means that a paired couple are not likely to compete directly with each other.

A good example of resource partitioning is provided by the red-eyed vireo, males and females of which exploit different parts of their habitat in the search for food.

The males hunt for insects at a greater height above the ground (on average around 36 feet) than the females (on average 13 feet). Because the ranges of both sexes overlap by only 35 percent or so, their feeding is largely non-competitive.

Resource partitioning between the sexes of a species does not only occur as a result of different behavior patterns, however. The male Arizona woodpecker has a much larger bill than the female, which allows him to forage on the trunks of trees, whereas the female searches for food in the branches.

Fighting for space
The main requirement of nesting penguins (left) is for space. Each individual maintains a hold on its own plot of territory by displays of aggression toward adjacent nesters. The result is a distinctively regular spacing among the population: individuals are separated from each other by two beak lengths.

Plants can also exhibit competition within a population, leading to a remarkably similar, regular pattern of distribution. In the Mojave desert, in North America, plant growth is restricted by the low availability of water and any area of land can support only a limited number of individuals. Of the many tickweed (above) seeds that germinate, only those that grow fastest obtain sufficient water.

SUCCESSION

When a lawn is left unmown, it changes. Not only do the grasses grow longer, but new, taller species invade because they are no longer excluded by mowing. So the composition of the community alters. This gradual change is called succession and happens in many different natural ecosystems. Succession may be initiated by a completely new environment when, for example, a glacier retreats, a shingle ridge is formed along a coast, or a volcanic island emerges from the sea. In such a situation the course of change is termed primary succession. Alternatively, an ecosystem may be severely damaged by catastrophe, such as lightning or fire, or by human activity. As the ecosystem heals itself and recovers, it is said to undergo secondary succession.

With both primary and secondary succession, the development moves in a series of fairly predictable stages toward a climax when the ecosystem reaches a natural equilibrium. While it is not always possible to say in advance what animals or plants will make up the climax community, the overall composition and structure of the vegetation is predictable. Often the climax community is determined by the prevailing climate, but other factors, such as geology, hydrology, or human land use, may modify the outcome.

Primary succession is the process by which a stable, mature, self-perpetuating ecosystem develops from a virgin environment. When sand dunes form above the high tide mark on a sandy shore, for example, the initial "soil" is saline and contains very little organic matter apart from the detritus of the tideline. It also quickly dries out in the sun and blows around easily in the wind. The result is a very inhospitable habitat for plant and animal life. Yet some plants can occupy this unlikely material, such as the sand couch grass (*Elytrigia juncea*), which can tolerate regular soaking by sea-water and can germinate and establish itself in unstable sand. As soon as this grass begins to grow it changes the physical conditions of its immediate environment. The basal sand is bound by its roots, and its shoots create eddies in the sand-bearing winds that lead to the deposition of sand grains around them. Soon small heaps of sand develop, only a few inches high but sufficient to elevate the surface a little above the zone of frequent flooding.

The new topography thus created opens up new opportunities to other species too sensitive to frequent flooding to cope as initial pioneers. Most influential among these is the marram grass (*Ammophila arenaria*), a much taller and more robust plant than the sand couch, which also soon begins to develop mounds of sand around its tussocks. Because of its size and vigor, the marram overgrows the couch, which ultimately succumbs to the competition as the dune continues to grow. This process is characteristic of succession, and it consists of two stages, facilitation and

Glacial retreat
The retreat of glaciers exposes pristine areas of ground on which primary succession takes place. The smooth rocks, polished by the passage of ice, are invaded by lichens and some of these contain nitrogen-fixing blue-green bacteria (cyanobacteria). They also secrete acids that attack and degrade the rocks. The flaking of lichens from the rocks and the grazing of small invertebrates upon them leads to a build-up of organic matter in rock crevices, where ground-up rock detritus may also accumulate, producing a primitive soil. In these crevices, a number of plant species, including woody plants such as dwarf willows, are able to gain a root-hold.

Eventually, taller shrubs and trees such as alder, birch, pine, and spruce are able to invade as the soil cover deepens and the level of organic matter increases. This provides a reliable source of water in the summer and the stability needed by the trees to cope with high winds in winter. Coniferous forest often forms the climax vegetation of this succession in the boreal and alpine regions.

Plant succession after the retreat of a glacier

Ice and bare rock with lichens	Herbs such as mountain sorrel growing in soil pockets	Willows and other dwarf shrubs form cover over bare rocks, and root in crevices	Least willow and mosses growing under and around snow patch	Ground layer of whortle-berries	Gray alder Aspen Birch	Scots pine	Norway spruce

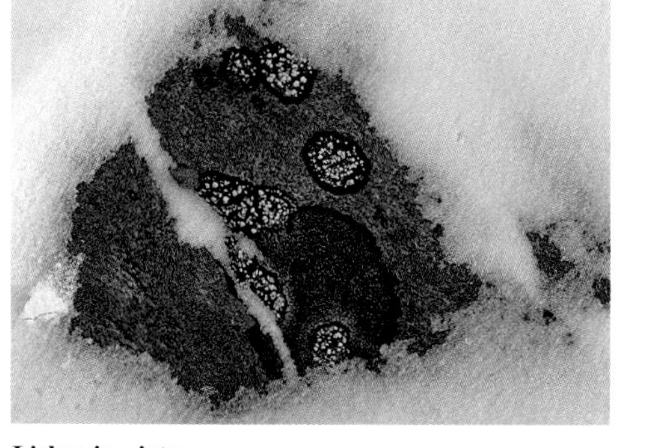
Lichen in winter

Mountain sorrel *Oxyria digyna*

competition. In facilitation, the presence of the initial colonist changes conditions enabling further immigrants to move in, just as the couch creates hillocks that protect the marram from flooding. The competition is when the initial colonist is ousted by the species it has inadvertently assisted. In this way, the pioneer species has ensured its own extinction.

As time proceeds in the sand dune succession, other species invade because the conditions become increasingly less inhospitable and extreme. The growth of vegetation not only stabilizes the soil, but adds organic matter that helps it retain moisture and allows the development of a decomposer food web. Nitrogen-fixing bacteria then arrive and nutrient cycles begin to operate. The shade of the tallgrasses provides a new microclimate in which mosses and more drought-sensitive herbs and invertebrates can survive, and these beds of moss provide a suitable environment for the seeds of shrubs and trees. Ultimately, the process leads towards a forest, the nature of which depends on the climate.

Although this type of succession is driven by the plants, outside factors also play a part. Without an external supply of sand the dunes could not have been built, and the immigration of some species, such as the trees, may depend on the chance arrival of a bird carrying the appropriate seed. So in some respects the succession is predictable (deterministic) and in others chancy (stochastic).

Successions can begin from a variety of starting points, such as the open water of lakes and ponds or the volcanic ash that follows an eruption, but the end point, or the climax, is reached when stability is achieved. Different successions take different routes but often converge toward a similar climax that is determined largely by the local climate. If a secondary succession is initiated by, for example, a natural catastrophe, what remains of the soil and the few surviving animals and plants (even if the vegetation cover has been completely stripped), help the community to quickly reach a climax again.

Despite the wide variety of forms that succession can take, there are some features common to all. Biomass generally increases to a maximum at climax, the physical conditions become less extreme and harsh, and the reservoirs of nutrients in the biomass and in the soil increase, leading to a degree of robustness and stability in the ecosystem. It is not always easy to tell when the climax has been reached: perhaps the key factor is how well the system can withstand disturbance without collapsing; also important is how fast the ecosystem can heal following a disturbance – its resilience. However, mature ecosystems are usually more complex in their structure (that is, in the architecture of their vegetation) and microhabitats. This increases the opportunities for a greater diversity of species to survive, leading to an increase in species richness during the course of succession. This complexity can mean that mature ecosystems are easily damaged, as in the tropical rain forest. Damage to the forest structure can lead to the loss of certain specialist species, and every species loss will affect the populations of other species dependent on it, so a cascade of extinctions begins.

Coniferous forest at the base of Herbert Glacier, Alaska

ISLANDS

The huge fruit of the coconut can float thousands of miles to colonize very remote tropical islands

Islands are found in almost every climate and every part of the world. But, whether tiny islets in tropical seas or vast icy islands like Greenland, islands provide a very special ecological environment because they are difficult to reach and colonize. This means they allow species to develop in isolation. They may have fewer species of animal and plant than mainland regions, but these species are often unique, or differ markedly from their mainland counterparts. Some species that might be driven to extinction by competition on the mainland may thrive on an island. Others develop in a way that is peculiarly suited to the island environment.

The most striking feature of island ecology is its lack of diversity. An island almost invariably has fewer species of plants and animals than its mainland equivalent because the sea acts as a barrier to normal dispersion. It may still have species not found on the mainland, but will usually have far less variety.

However, the sea is by no means an insurmountable barrier, and many plants and animals are able to reach islands, particularly large ones or those near the mainland. Few islands would have much life at all if they were not colonized from the mainland. Sea birds and sea mammals such as seals have no problem crossing over. Some coastal plants, too, are ready colonists of islands, especially those whose seeds are spread by waves. Coconuts are regularly swept thousands of miles across the ocean to germinate on isolated tropical islands. Coconuts from the Caribbean are often washed up on the west coast of Ireland, but the climate there is too bleak for them to grow.

Many plants produce fruit light enough to be windblown across wide stretches of ocean to invade islands. Other seeds are carried in the guts of migratory birds. Land mammals, birds, and many plants, however, only rarely find their way onto isolated islands, and their chances of survival and successful breeding are slim. Even so, quite large rafts of vegetation can float for many days in the ocean, covering hundreds of miles and carrying many species, including animals and plants, to islands.

Once an organism has arrived at an island and established a breeding population, it may find that it has no natural predators and few competitors on the island. This reduces the pressure of natural selection, and a once-specialized species may take up a more general role in the community and develop a wider niche.

However, there may be natural physical barriers on the island, such as mountain ranges and wetlands, or neighboring islands that can be more easily colonized but present some barrier to interbreeding within the population. When this is the case, the colonist species may develop along different lines within each of its isolated populations and radiate into a number of separate species – as with Darwin's finches in the Galapagos Islands.

Those species that disperse easily need not dominate island populations. Since most individuals that leave the island will perish, natural selection favors species with mechanisms that prevent them from

New Zealand's flightless birds
New Zealand's large size, lack of native predators and isolation led to the evolution of several flightless birds. Giant moas are now extinct, but three species of kiwi (left) survive. The kakapo (right) is a flightless parrot now driven close to extinction by introduced predators and human disturbance of its habitat. The takahe, a striking flightless gallinule, was believed to be extinct until its rediscovery in 1948, and it survives in small numbers on the South Island.

Krakatau
In 1883, the Pacific island of Krakatau exploded in a volcanic eruption of such enormous force that the noise of the explosion burst the eardrums of sailors over 25 miles away. All that remained after the explosion was a ring of smaller islands, completely devoid of life. Three years later, in 1886, biologists began to monitor the recolonization of one of these lifeless islands, Rakatau (right), and we now have records detailing the changes over the past century. As the graphs below show, the numbers of plants and animals are still increasing, but the rate of increase for most types of organisms has now slowed down considerably.

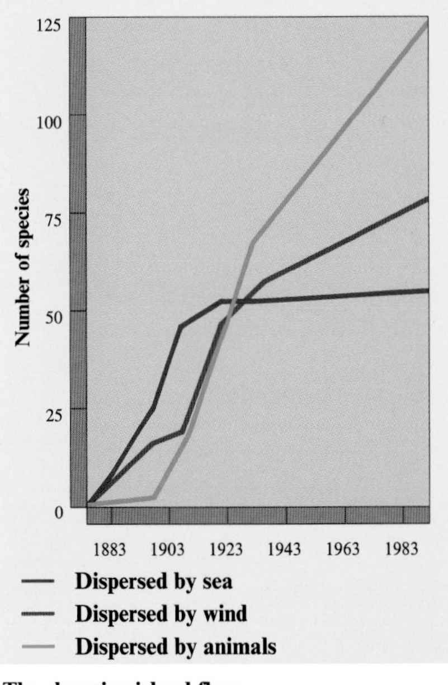

The changing island flora

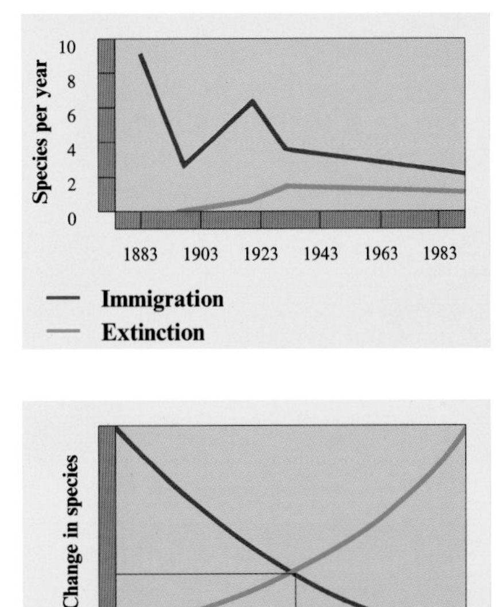

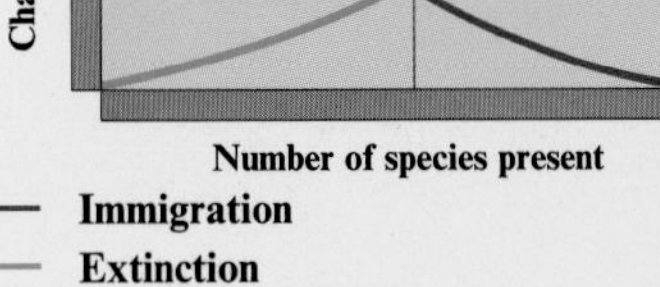

Island biogeographic theory
The number of species an island can support is affected by two main factors: its size and its distance from sources of colonizers. Large islands or those near sources of colonizers tend to have the most species. Small remote islands have the fewest species. But the more species there are, the more competition there is, and the more are driven to extinction.

The number of species is thought to stabilize when the rate of immigration of new species equals the rate of extinction of existing ones. The immigration rate may be high on an empty island, but declines as the island fills up with species because there are fewer appropriate species available for further colonization. Conversely, the extinction rate becomes higher as the island becomes more crowded with species. The point where these two curves cross on the accompanying graph gives the equilibrium species number. On a large island the extinction rate will be slower so a higher equilibrium number is to be expected.

leaving. So many land birds and insects that inhabit remote islands are flightless.

Animals and plants on islands often change size in response to reduced competition and predation. On the Mediterranean islands, there are fossil remains of pygmy elephants that once lived there. These probably evolved from large animals because the limited resources of the islands could better support smaller animals. But where there are no native trees, smaller plants may evolve into giant forms that can fulfill this ecological role. On the Hawaiian Islands, the major endemic tree species belong to the daisy family (*Asteraceae*). Giant animals are also known on islands, such as the Komodo dragon, a massive predatory lizard able to kill water buffalo. The extinct dodo of Mauritius was an overgrown flightless pigeon.

THE URBAN ECOSYSTEM

Whereas most of the world's ecosystems are in retreat, urban habitats are expanding in virtually every continent. Even if there were to be no further growth in the planet's human population, urban habitats would still go on expanding because of the constant drift of people from the countryside to the towns and cities. Urbanization has had an even greater effect on the world than crop cultivation. Even as recently as 1900 only one country, Great Britain, could be said to be urbanized. Yet it took only 90 years before half the world's human population was living in cities, and barely 10 more may see this rise to 75 percent.

Urban areas differ from their surroundings both in major factors such as climate, soils and the amount of pollution present, and in more subtle ways through the myriad effects of human occupation. The ecosystems prevalent in urban environments can nonetheless be said to possess unifying characteristics.

Cities are very new environments. Even the oldest is no more than a few thousand years old. Most are less than a century. So the species that now make their home in the cities all originated in different habitats. But often their urban homes bear a marked resemblance to their natural habitat. Ledges, walls, and friezes of high buildings are not so very different from cliff edges. Cliff-nesting birds such as the kestrel, gulls, the common swift, the town pigeon (which evolved as the rock dove) and the black redstart have quite happily extended their range to include tall buildings and pylons, on which they are as at home as on the ledges of more natural cliffs.

Similarly, for bats and birds such as swifts attic spaces and ventilator shafts clearly bear some resemblance to their native caves and holes. The same may be true for the Norway (brown) rats which have thrived in European city sewers and drains since Roman times – and robustly defied countless attempts to oust them. These rats once lived in holes in river banks in Southeastern Asia (not Norway), but traveled in ships' cargoes to the cities of Europe and many other countries where they bred rapidly in the sewers, feeding on rubbish and virtually anything. Brown rats, along with house mice (also natives of Southeastern Asia), have adopted the city life so thoroughly that they are now genuinely urban species. It now seems that they may even be evolving resistance to the warfarin poison developed to eradicate them.

Like rats and mice, some species have found the opportunities in cities so attractive that they actually prefer them to their natural habitats. Starlings, for example, now regularly roost in huge numbers in city centres, wheeling in at sunset in vast, noisy flocks after a day feeding in the surrounding countryside or suburban gardens. The warmth of the city at night, and the abundant ledges make the city a perfect dormitory for these birds.

One potent attraction of the city is the vast amount of organic rubbish generated. Many kinds of scavenging creatures have moved in to take advantage of the ready-made feasts in the cities as their own natural habitats have shrunk. Jackals and hyenas often sneak into Indian and African cities, while coyotes are sometimes seen in the outskirts of North American cities. In the Arctic, even polar bears make occasional visits.

Other creatures may be drawn by the warmth of the city. The heat generated by garbage dumps is good for lizards, slow-worms and crickets, while the ice-free water in many city rivers and canals provides extra feeding opportunities for birds in winter.

Opportunist species

It takes time for birds and mammals to get round to occupying the new ecological niches created by urbanization. The first ones to manage are those that are quickest to discern opportunities and to make the best of them. The house sparrow has been phenomenally successful on both sides of the Atlantic. Its secret seems to be that it is a non-specialist feeder, it is flexible in the choice of nesting sites, it is very tolerant of disturbance, and it has a high level of intelligence (shown in maze tests to be comparable to that of rats and monkeys). Other opportunist birds, possessing something of the same range of aptitudes, include the magpie, the crow, the jay, and the herring-gull.

Most surprising of all, perhaps, is the red fox. Foxes are occasionally seen in cities such as New York, Toronto, Amsterdam, Paris and Stockholm. But they have become almost a permanent feature in British cities. As cities have grown, and feeding opportunities in the country have contracted, foxes seem to

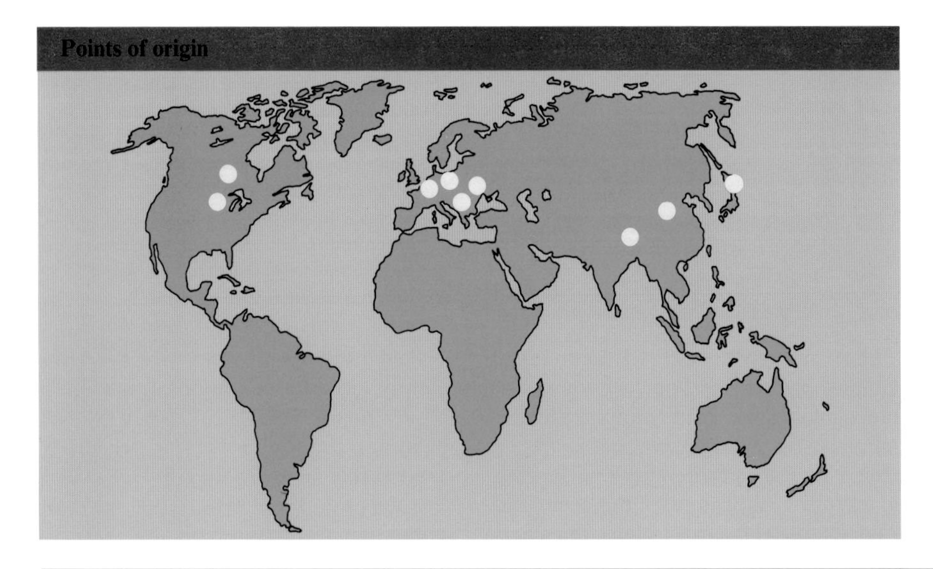

Wasteland immigrants

Plants colonize urban wasteland surprisingly quickly. Among the earliest arrivals are often introduced shrubs and herbs, such as rosebay willowherb, which seem to thrive in the tough conditions. In British cities, for example, a single patch of wasteground may provide a home to: sycamore, laburnum, wormwood, and goat's rue from Europe; buddleia, Japanese knotweed, and *Cotoneaster microphyllus* from the Far East; and Canadian goldenrod and Michaelmas daisy from North America. These plants draw to the wasteground butterflies and bees, and birds such as linnets and greenfinches. Wherever rosebay willowherbs thrive, so often does the elephant hawkmoth, whose caterpillars feed upon the leaves.

have moved into towns with remarkable success. They are not primarily scavengers, but feed off bird tables or compost heaps, and dig for earthworms in city gardens, where they often live. Every year up to 60 percent of city foxes die, though, mainly killed by road traffic, and this has an inevitably major effect on the stability of family groups.

Perhaps the most prominent of all urban hunters are domestic cats, which may be either pets or strays. But analysis of their average stomach contents has shown that their diet is obtained partly by raiding trash cans and refuse tips, and partly from food put out by cat-lovers. as well as by catching birds and mice.

Of all the creatures that live in cities, the most successful, perhaps, are the least conspicuous – the billions upon billions of invertebrates that find their way into every little urban nook and cranny. Cockroaches, crickets, and spiders revel in the warmth of buildings, while flies feast on huge quantities of dog feces, and parasites such as fleas, lice, and bedbugs find plenty of unwilling hosts. In the tropics, an urban species of mosquito has evolved, *Aedes aegypti*, which transmits yellow fever. In many city houses, there are as many species of insects as in a substantial area of woodland.

Exotic species

Countless foreign organisms are introduced, both intentionally and unintentionally, each year into the cities. Many of them eventually find a vacant ecological niche and become more or less permanently established. Plants are the most numerous and most important exotic organisms in this respect, for they are to some extent matched with the local climate.

Towns are accordingly focuses for the initial introduction of populations of plant species that thrive under disturbed conditions, and that later spread into the countryside, as European sycamores and rhododendrons have done. Studies in Berlin have shown that there is a close correlation between the growth of the human population and the number of introduced plant species that escape into the wild. Some 50 percent of the flora of central Berlin are introduced species ready to "migrate," or that have spread already.

The typical pattern of spread begins rather slowly and uncertainly, but eventually picks up speed, triggered more often than not by changes in land use. During World War II, for example, the multiple creation of bomb-sites in many European cities enabled many species to gain a firm foothold for the first time. On a geological timescale, most garden plant species have only just arrived – so many more naturalizations should be expected.

Animal introductions, too, have often proved surprisingly successful in urban areas. For example, the Common mynah, a starling-like bird from India, is now virtually a native of cities in South Africa and Australia.

Garbage feeders

Many creatures are drawn into cities by garbage. Rubbish tips in tropical cities make popular feeding sites for maribou storks and black vultures, while tips in colder climes attract gulls – especially herring and black-headed gulls. Small mammals like raccoons, opossums, and even badgers may visit urban trash cans. In the tropics, scavenging baboons have become a real problem.

HABITAT

Ecosystems

1 Taiga 28-31

2 Tundra and Polar 36-39

3 Oceans 44-47

4 Coastlines 52-55

5 Swamps and Mangroves 60-63

6 Temperate Grassland 68-71

7 Temperate Forest 76-79

8 Savanna 84-87

9 Rain Forest 92-95

10 Mountains 100-103

11 Rivers and Lakes 108-111

12 Deserts 116-119

13 Scrublands 124-127

14 Corals 132-135

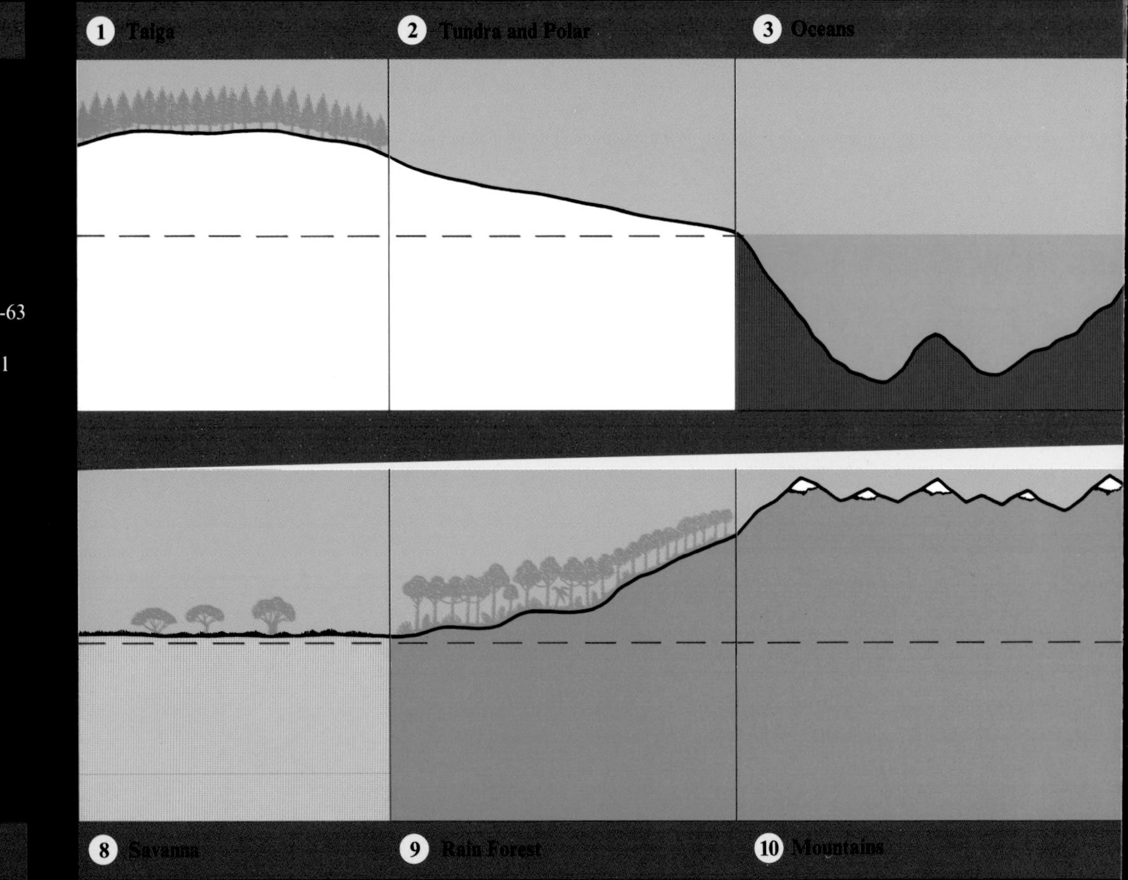

S IN FOCUS

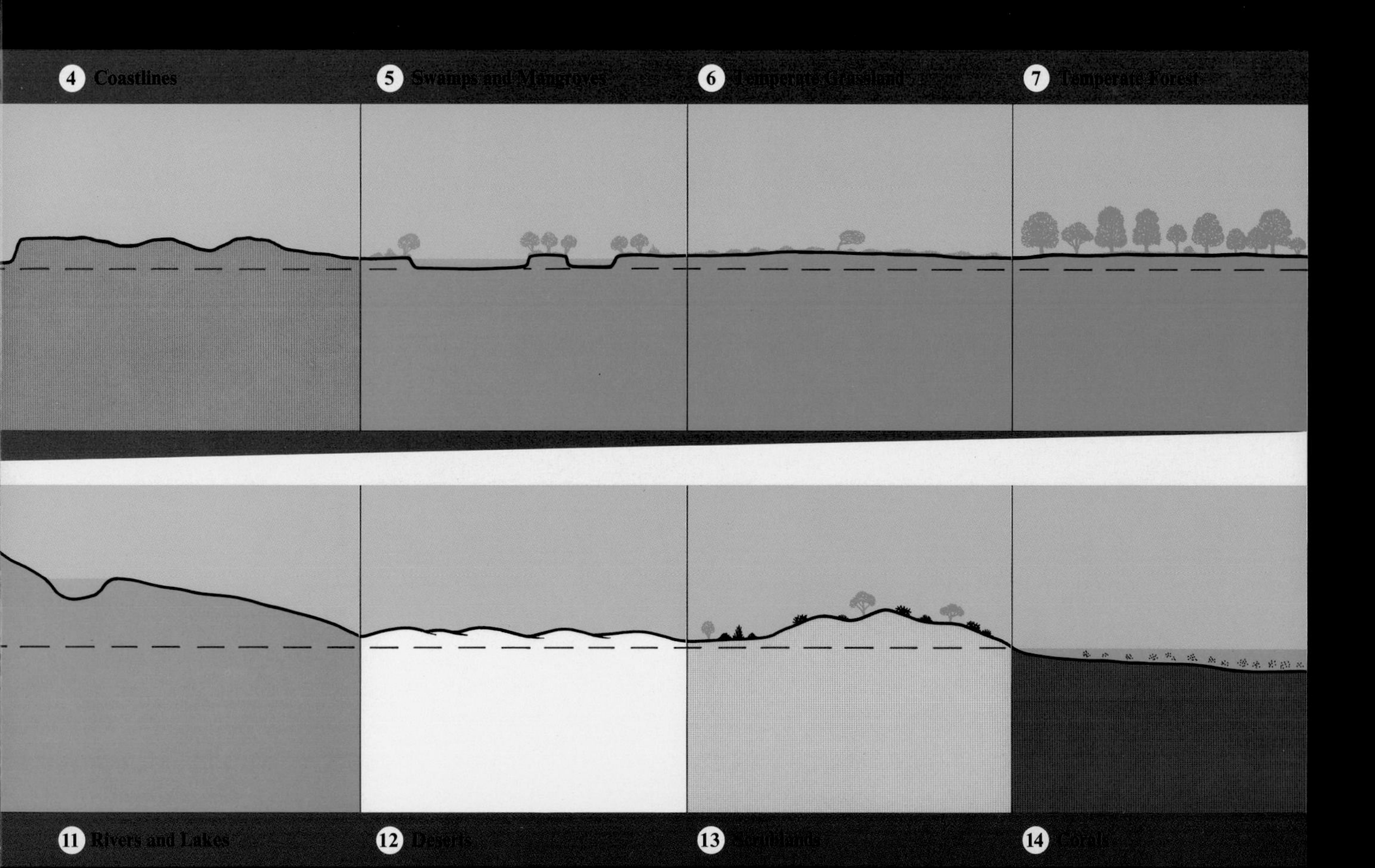
4 Coastlines
5 Swamps and Mangroves
6 Temperate Grassland
7 Temperate Forest
11 Rivers and Lakes
12 Deserts
13 Scrublands
14 Corals

TAIGA

Vying with deserts as the largest of the world's biomes, the great northern coniferous forest stretches almost 6,800 miles around the top of the globe. It is known as the boreal forest (from Boreas, Greek god of the north wind) or taiga (from the Russian word for a marshy pine forest). The boreal forest covers about 11 percent of the Earth's land surface. There are a few scattered remnants in Scotland, then it runs in a virtually unbroken belt right across northern Eurasia, from Scandinavia to the Pacific shores of Siberia and on to northern Japan. In North America, it clothes much of Alaska and Canada, extending south into New England. Similar forest is found on high mountains at lower latitudes, such as the Alps in Europe or the Appalachians or the southern Rockies in the USA.

The northern limit of the boreal forest corresponds approximately to the southern extent of the Arctic front in summer, and its southern limit to the winter extent of the Arctic front, which generally lies around latitude 58°N. There is no equivalent forest in the southern hemisphere because the latitudes at which it would grow are occupied by ocean.

The taiga is bordered to the north by the tundra, where no trees other than dwarf willows can grow because the summers are so brief. In the taiga there are generally 30 days or more when the light is sufficient and the temperature rises high enough (50°F or more) for full-size trees to grow. As with other transitional zones (ecotones) between biomes, there is not a neat, abrupt change from bleak treeless tundra to dense dark conifer forest. High, exposed ground and swampy regions within the taiga where trees cannot grow contain isolated islands of tundra, while tongues of taiga penetrate deep into the tundra in sheltered river valleys and hollows.

Moving south from the tundra proper, there is a zone of forest tundra ranging from a few miles to hundreds of miles wide. Here, areas of stunted birch, alder, and willow trees occupying sheltered pockets are interspersed with lichens, mosses, and grasses. Farther south scattered stands of tall conifers appear, then open boreal woodland. In North America, for example, this region consists chiefly of black spruce and lichens. Still farther south comes the boreal forest proper, with huge areas occupied by continuous stands of spruces, pines, firs, and larches, broken here and there by glacial lakes, rivers and streams, bogs, and thickets of alder and other deciduous trees. To the south of the boreal forest lies either mixed or pure broadleaved forest or temperate grassland.

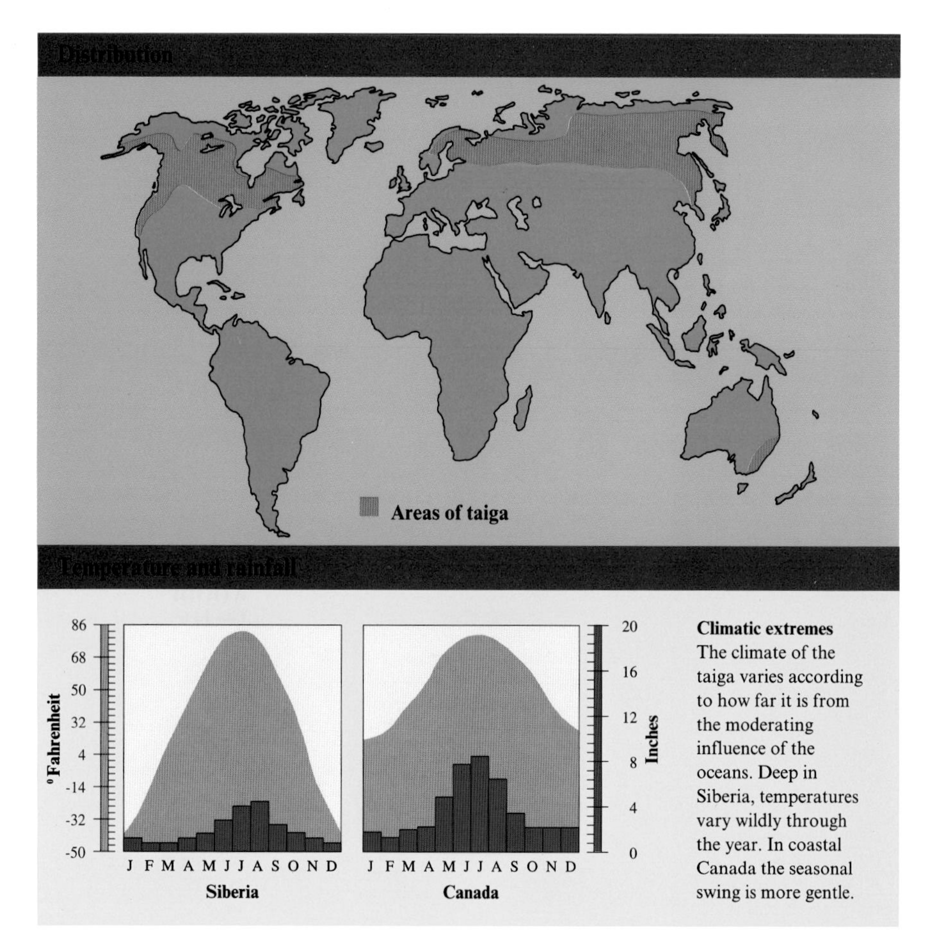

Climatic extremes The climate of the taiga varies according to how far it is from the moderating influence of the oceans. Deep in Siberia, temperatures vary wildly through the year. In coastal Canada the seasonal swing is more gentle.

Climate

Although there is enough light and warmth for tree growth in the boreal forest, the climate is still harsh. In winter, raging blizzards pile snow into huge drifts that stay unmelted for more than six months of the year. Water is locked up for much of the year in ice and snow, so that the vegetation has to cope with drought. Yet because little moisture evaporates in this cold climate, there may be too much water when the spring thaw comes. Then decomposition is hindered by waterlogging and plant remains accumulate as peat. Soils are often acid and relatively infertile. If they are well drained the nutrients are often leached out and locked up in a hard layer deep within the soil.

Temperatures in the boreal forests can be lower than on the tundra in winter, especially in the interior of the continents. In the central areas of Alaska and Siberia, far from the moderating influence of the sea, the seasonal temperature fluctuations are the largest in the world. In the Verkhoyansk region of central Siberia, for example, winter temperatures are the coldest on Earth outside Antarctica, dropping to as low as –90°F, while in summer they may soar to well over 85°F – a staggering difference of over 175°F .

Productivity and diversity

Because of the short growing season, productivity within this habitat is limited. Indeed, its yearly average is only about half that of temperate deciduous forest, for a similar biomass.

Compared with forested habitats to the south, there is much less diversity of plants and animals. This may be a result not only of the harsh conditions but of the relative youth of the boreal forest

The taiga, or boreal forest regions of the world, are covered in frost for up to 10 months of the year

Most of the taiga is waterlogged because permanent underground frost prevents any surface water from draining away

habitat. Whereas the glaciers had largely retreated from the areas now occupied by temperate forests by about 14,000 years ago, parts of the boreal forest region were heavily glaciated until quite recently – and some areas within the region, for example in Alaska and Norway, still contain large glaciers.

Despite this relative uniformity and the vast expanses occupied by only one or two species of trees, the boreal forest is not always the monotonous blanket it might at first seem to be. Variations in the permafrost layer, altitude, aspect, drainage, and other factors all produce subtle variations in the vegetation. In North America, for example, cold, damp, north-facing slopes and hollows are colonized by black spruce, which can survive waterlogged soil and shortages of nutrients; the drier, warmer sites found at higher altitudes, and also in regions regularly cleared by fires, are occupied by Jack pine; and south-facing slopes without a permafrost layer are the home of white spruce and birches.

In Russia, too, there are variations across the great expanse of the taiga. The western part, with a moister, significantly warmer climate, is the province of the spruces. In the harsher conditions of the vast central region of Siberia, and along the northern border with the tundra, these are replaced by the hardiest of conifers, the larches. In the far east, and in the mountains, the Siberian stone pine is the dominant tree. Mixed in with these are other trees – for example, birches quickly take advantage of areas where the conifers have been cleared by logging or fire, while Scotch pine grows in areas of sandy soil.

Animals

The animal life of the boreal forest is limited by the small number of different niches available and by the severity of the winters. The major large herbivores are the deer, with more species found here than in any other biome, while small herbivores are represented chiefly by rodents, from voles to beavers. Predators include numerous species in the weasel family, as well as lynxes and wolves. Resident birds include several species of grouse. Many of the region's birds are summer visitors only, migrating to temperate regions or farther south for the winter. Few amphibians and reptiles are able to survive the cold winters; all the common taiga reptiles give birth to live young because there is too little sun to warm eggs. Insect life, on the other hand, is far richer than in the tundra.

In addition to being clothed in a thick insulating coat of fur or feathers, which is especially luxurious in winter, many of the animals of the northern forests are particularly large compared with relatives in kinder climates – an advantage when it comes to conserving heat (the larger an animal, the smaller is its surface area, from which heat can be lost, relative to its bulk). For example, the elk (moose) is the biggest member of the deer family, the wolverine the largest of the world's weasels, and the capercaillie the biggest of all grouse.

TAIGA DYNAMICS

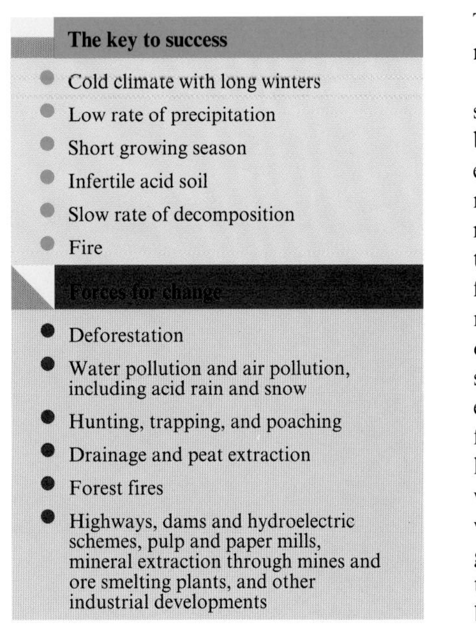

The key to success

- Cold climate with long winters
- Low rate of precipitation
- Short growing season
- Infertile acid soil
- Slow rate of decomposition
- Fire

Forces for change

- Deforestation
- Water pollution and air pollution, including acid rain and snow
- Hunting, trapping, and poaching
- Drainage and peat extraction
- Forest fires
- Highways, dams and hydroelectric schemes, pulp and paper mills, mineral extraction through mines and ore smelting plants, and other industrial developments

Bitterly cold in winter, and often far from the world's major population centers, the boreal forests of the north have long avoided the widespread human devastation that has affected so many ecosystems. Yet they have not escaped entirely, and the impact of human activity is increasing – especially in Siberia.

The taiga's resources were exploited in a sustainable way for thousands of years by Native Peoples, both in Eurasia and North America. Yet as long ago as the late 1600s things began to change for the worse, when rising demand in Europe sent explorers and fur traders pushing northward to hunt the taiga's wealth of fur-bearing mammals – foxes, minks, sables, and beavers. Bitter wars were fought over the fur trade and many people died. But it was the animals that were, as they still are, the losers. As demand for the pelts grew dramatically in the 19th and 20th centuries, huge numbers of these creatures have been shot or trapped and slaughtered.

The logging trade

The next stage in the exploitation of the taiga was logging. The boreal forest is the world's richest source of softwood timber and pulpwood for making paper and other materials, and over the last two centuries it has been increasingly exploited. In some areas, such as Sweden and Finland, large-scale logging began early – as long ago as the 17th century in Sweden, where the timber was used industrially as fuel in smelting rich deposits of iron ores to obtain metals. Today, precious little primary virgin forest remains over much of Fennoscandia.

Siberia, Russia, and North America still contain vast areas of boreal forest, but here, too, there has been a history of exploitation with little or no attempt at regeneration – a process that can take many centuries. Less than 30 percent of the 7.5 million acres of boreal forest felled each year in Russia and Siberia is reafforested. In many areas, what was once forest is now bog, and felling on slopes has inevitably resulted in soil erosion. Moreover, rivers have been filled with millions of logs, so that they have silted up. The result is that the water has become poisoned and fish and water-birds have been affected. Altogether, an area of northern Asia greater than France has been severely affected by the activities of the loggers.

Mining in the taiga

Large-scale mining for iron ore, gold, and other minerals has also left its mark on significant areas of taiga. Russia is one of the world's major gold producers while a single source of diamonds in Siberia yields almost a quarter (by weight) of the total world output. As well as the damage done by the mining itself, the roads, settlements, and other infrastructure have totally disfigured the natural environment in places. Worse still, ore smelting plants, pulp and paper mills and other industrial plants have produced massive air pollution in Canada and Russia.

Vast hydroelectric schemes, such as that on the James Bay in Canada and the Angara River in Siberia, are drowning huge areas of taiga with unpredictable consequences for Native northern communities, who see their trapping grounds disappear beneath the water. The mining of peat to supply fuel for power stations and for the horticultural trade, together with the drainage of bogs and other wetlands, adds to the toll of destruction. Financially lucrative oil and natural gas extraction are becoming increasingly important, too – with potentially serious implications.

Air pollution

For some time, the taiga seemed cushioned from the pollution that has afflicted so much of the industrialized world. But there are signs that the spread of industry in the north of Russia and elsewhere is bringing poisonous air and water to this once pristine environment. Moreover, trees in the boreal forest are now being affected by the acid rain generated by the burning of fossil fuels in industrialized areas of Europe and North America far south of the taiga. Acid rain causes both direct damage to the trees and weakens them, allowing insects and other pests to attack more easily. It also plays havoc with fish populations and other wildlife.

Twigs are dragged to the beaver's food store

One serious "hidden" effect of acid rain is its ability to dissolve metals that were previously locked up in the soil. These liberated metals can be seriously toxic to tree roots and soil microorganisms. One liberated metal, aluminum, irritates the skins of fishes when washed into a stream, until eventually their gills clog up with mucus and they

analog in the north: the stubby-winged, fish-eating guillemot.

The permanent residents of the tundra have evolved various strategies to cope with the cold and wind. Most tundra dwellers have compact bodies with short limbs, bills, and wings to conserve heat. Many are also insulated from the cold with thick subcutaneous deposits of fat or blubber, or dense layers of feathers or fur that trap air next to the skin. The main, protective guard hair of a fur seal may have 50 or more secondary soft hairs growing from the same follicle to trap extra air.

Wherever possible, animals burrow into the snow to avoid the hostile winds and to utilize the insulating properties of the earth, or even of snow and ice (Inuit igloos made from ice blocks can become very warm inside). When this is not possible, cooperative behavioral strategies may have to be developed. Emperor penguins living on the rocky coasts and islands of Antarctica huddle together in groups of several thousand to keep each other warm, constantly circulating and taking it in turns to spend some time on the exposed fringes of the group before returning to the communal heat of the interior. These birds do not build nests, but balance their eggs on their feet to keep them off the cold ground, and completely surrounded by warm living skin.

Cold-blooded creatures have their own problems at these extremes, and their own ingenious solutions. Around Antarctica, in particular, much of the life is in the sea, and the *Trematomus* fishes have evolved a protein compound in their blood that acts as anti-freeze, preventing ice-crystals from forming and allowing the fish to survive temperatures below 28°F. Some species of insect also contain a type of anti-freeze in their bodies, while Alaskan midge larvae can be frozen and thawed over and over without apparent damage.

Polar winter
The sun barely rises during the polar winter, and polar creatures such as the penguin and polar bear need thick layers of fat and fur to insulate them against the extreme cold.

polar summers, the tern
nds its life in almost constant
light. The migrants attract
ny predatory birds, like eagles
C), falcons and merlins.
Most leave again with the
arture of their prey, but a few
l species remain throughout
year, such as the rock
rmigan which feeds on seeds
l berries remaining under the
w, and the snowy owl (7C)
ich hunts the ptarmigan,
mings and hares.

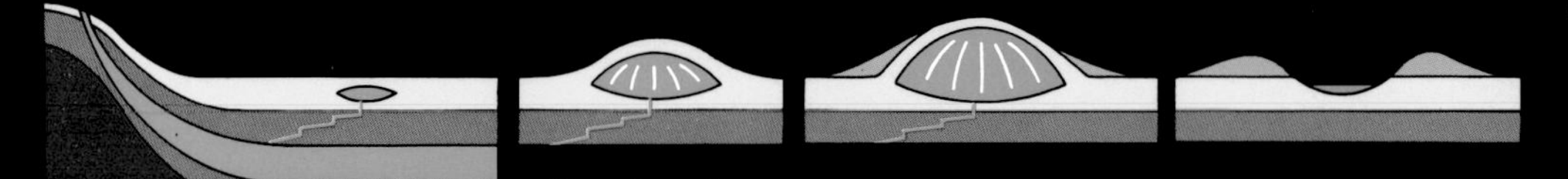

Pingo formation

LANDFORMS
The tundra is studded with humps of earth called pingoes (12A), which form when a crack grows in one of the impermeable rocks sandwiching a water-bearing layer. Under pressure from gravity, water rises up through the crack and it freezes as it nears the surface, forming an expanding ice lens. If the ice melts, the mound collapses into a pond bordered by slippage material. Such ponds provide homes for insects and water birds.

The huge seasonal temperature range causes strange patterns in the ground, called polygons (7D), when water seeps into cracks and levers patches of earth free as it expands and contracts during repeated freezing and melting.

17 | 18 | 19 | 20

A

B

C

D

16 | 17 | 18 | 19 | 20

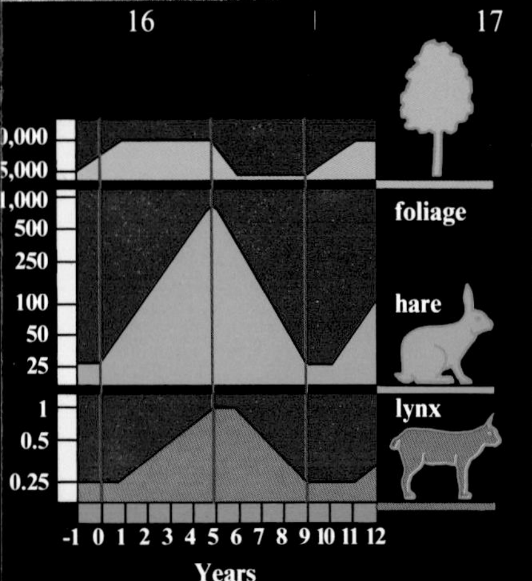

POPULATION CYCLES
The populations of the snowshoe hare (10C) and the lynx (5D) are intimately linked, and are connected to the amount of young vegetation available to the hare to browse.

Each hare needs approximately 10 oz a day of young growth, and excessive browsing or girdling – where a ring is eaten completely around a plant stem, cutting the transport of water and nutrients and killing the stem – can reduce the palatable vegetation well below this level, even though there is in fact plenty of vegetable matter around. Many of the dwarf birches and willows that the hare might be expected to eat secrete resins and other nauseating chemicals to discourage them. The resulting decline in the hare population is accelerated by predation pressure from the lynxes, whose numbers have not yet begun to decline and only start to do so when they have decimated most of the remaining hares. Suddenly, with the underbrowsed vegetation on the increase and with a depleted population of lynxes to contend with, the hare population can begin to rise sharply. Each of these populations undergoes a remarkably consistent four- to five-year cycle of rise and fall, with each cycle lagging slightly behind the one before, as the consequences of a lack of foodstuff take time to work their way up the food chain.

is that, during winter, the willow will be completely covered by an insulating layer of snow.

Any individual plant is made up of subpopulations of leaves, petals, and buds, each at a different angle to the Sun, absorbing different amounts of light and creating different amounts of shade. Many Arctic plants have fine hairs on their stems, which trap a warming layer of static air. Yet the roots may penetrate to temperatures only just above freezing in the space of a few inches below the surface. As a result the thermal map of an Arctic poppy (3D), for example, will show a wide range of temperatures over a tiny distance.

In this incredible variety of microclimates, some parts of the plant may succumb to the conditions while other parts, and the plant as a whole, continue to thrive. Many Arctic plants are protected by frost-hardening during winter, when their cells toughen and become resistant to dehydration.

A common tactic in the summer, used for example by the Arctic poppy, is to track the course of the Sun, with the petals of the flower acting as parabolic reflectors to concentrate light and heat on the seeds, which ripen more quickly as a consequence. Opportunistic insects will often keep themselves warm by hitching a ride in these flowers, and in the process may play a role in pollinating the plant.

BIRD MIGRATIONS

The presence of large numbers of invertebrates in the lakes, streams, and peaty, waterlogged ground attracts close to a hundred different species of bird to the area to breed, including many wading birds, ducks, and geese.

The willow grouse (10D) makes only a small journey from the boreal forests, but the Arctic tern (16A), the greatest migrant on Earth, travels 25,000 miles twice a year, from the Arctic to the Antarctic and back again. Because of the continuous days of

11 | 12 | 13 | 14 | 15

11 | 12 | 13 | 14 | 15

HUNTING AND HIDING

The most fearsome predator in the Arctic is the polar bear (14B), which haunts the coasts and ice floes, traveling up to 12 miles a day and preying largely on seals (17D). Even the heavily tusked walrus (18C) is not safe if caught unaware, although in both cases the bear is more likely to take cubs than adults.

In fact the tusks of the walrus are more for combat between the rival males than to deter predators, just as their tough, wrinkled hides and thick layers of fat are for defensive padding as much as to keep out the cold. The polar bear often lies in ambush for seals near airholes in the ice, or sneaks up on them, taking advantage of the camouflage provided by its snowy pelt.

The polar bear is always white, but other Arctic inhabitants such as the Arctic fox (11D), the hare (10C), and the willow grouse (10D) change color with the seasons, from various shades of gray or brown in the summer to a camouflaging pure white in winter.

UNDERGROUND HAVENS

Female polar bears dig burrows in the snow and and ice in which to have their cubs. Perhaps surprisingly, snow is an excellent insulating material, and although hibernation is not possible in the tundra, because the short summer does not allow the animals to put on enough reserve weight to last the long winter, many animals nevertheless burrow into the snow for shelter, including the hare and the Arctic lemming (11D). The Arctic fox even leaves buried larders of frozen dead animals to see it through the winter. Emerging from their burrows in the summer, the lemmings multiply rapidly. A lemming can be sexually active at 15 days old, and have its first litter in little over a month.

When there are too many lemmings for the local vegetation to support, they will migrate in their millions, often leaping into a gorge and swimming to the other side rather than taking a detour, a habit which has led to the tales of their mass suicides.

TUNDRA AND POLAR DYNAMICS

The key to success

- High atmospheric pressure keeps the temperature and precipitation low
- Herbivores should be less than the carrying capacity of the tundra vegetation
- Permafrost remains undisturbed
- Pollutant levels in fringing oceans lie within acceptable levels
- The stopping-off points of birds like the knot that migrate from pole to pole themselves remain undisturbed

Forces for change

- Uncontrolled culling of predators, as pests or for profit
- Uncontrolled civil engineering projects
- Pollution incidents (extensive and often disasterous oil leaks from pipelines, ocean transporters)
- Excessive harvesting of krill, fish, or seals
- Erosion of the delicate peaty soil by vehicles

Too cold for much permanent habitation, and far from the large industrial centers of the world, the tundra remained until recently one of the last unspoiled wildernesses of the world. Sadly, this situation is beginning to change, especially with the discovery of oil in northern Russia and Siberia.

Civil engineering works connected with oil exploration and extraction have disturbed the permafrost over many Arctic regions. One effect of this is to disturb and compact peaty surfaces, reducing the insulation from the cold provided by overlying soils and plant litter. The result is further melting of the permafrost, subsidence, and dramatic erosion. Pipeline construction, too, has melted the permafrost in places and can provide a major obstacle to the seasonal migration of large herbivores and other animals. Lines to guide engineers in taking seismic readings have also been bulldozed across the tundra at regular intervals, causing serious damage. Even where this does not lead to erosion, slow-growing tundra vegetation can take many years to recover.

Unlike many ecosystems, the tundra has suffered little from overgrazing in the past. Yet there is growing evidence that caribou numbers may be increasing, partly as there are now far fewer human hunters, but also because several of their natural predators, such as wolves, have been subjected to severe culling in some Arctic regions. If caribou numbers do increase significantly, they may well begin to destroy the slow-growing lichens which comprise some 50 percent of their winter and spring diet.

Pollution on the ground, north and south

Far from the sources of pollution as the Arctic and (particularly) the Antarctic are, the atmospheric transportation of pollutants is now global. Residues of pesticides and industrial effluents have been found in Canadian Arctic wildlife, and polychlorinated biphenyls (PCBs) in the fat of Antarctic penguins.

The nuclear accident at Chernobyl was responsible for severe radioactive pollution in the domestic reindeer population of Lapland through the long-lasting contamination of the lichens eaten by the animals, who were then slaughtered in huge numbers to prevent radioactive contamination of the herders.

Major oil-spills have occurred in Alaska and in the eastern Canadian Arctic, and, although the long-term effects may as yet be unknown, caused considerable local damage. In the Antarctic – where there is less seaborne pollution in general – the rubbish brought into the area by explorers and temporarily resident scientists takes so long to decompose that many settlements are marked by veritable mountains of imported garbâge, despite the international guidelines set down for the removal of all such wastes from the bases. Similar bases in the Arctic also suffer considerably from this problem.

Marine ecosystems

The human overexploitation of some of the edible species of the polar seas has had a significantly destabilizing effect on some ecosystems. Modern methods of fishing tend to result in the major reduction of local fish populations, inevitably thus also depleting the main food source of seals and whales, who then themselves dwindle in number. Small species of fish right down to tiny plankton are now regularly caught and processed on a virtually industrial basis, further disturbing the food web of the

The southern right whale

Few Antarctic creatures have suffered as badly at human hands as right whales, called "right" by whalers because they gave the longest whalebones and thickest blubber. Hunted close to extinction, the 1,500 or so right whales mate every winter off Patagonia's Valdes Peninsular in bays now threatened by oil and gas development.

normally productive oceans.

The ecology of the Antarctic is dominated by the sea. At the very hub of the food web of the Southern Ocean is the shrimplike crustacean called krill (*Euphasia superba*). This invertebrate occurs in vast numbers: it may measure a maximum of only 2.4 inches in length, but at its highest shoal density there may be up to 2 ounces of them for every square yard of sea. The krill swim in the top 150 feet of sea in formal schools. Fur seals, penguins, albatrosses, and several fish species are totally dependent on them for food.

Present concern centers on the fact that although the numbers of the krill have always varied from year to year, the current intensity of human fishing for them may lead to an overall decline in its abundance, which could affect very many more animals, notably the blue whale, which is also totally dependent on them – and which is already an endangered species.

Climate changes

The polar regions are highly vulnerable to changes in the climate because they depend for much of their ecological stability on the presence of ice and ice-related factors within the soil. Large areas in the polar north consist of tundra that has developed on peat and other organic-rich soils overlying permafrost. Any change in the environment that reduces the amount of ice in the soil – such as climatic warming – would irreversibly modify the ecosystem dependent upon it. It would also hasten the decomposition of the large store of organic matter in the Arctic soil, further increasing the carbon dioxide content of the atmosphere, and accelerating the "greenhouse effect." Although there are few comparable areas at the south pole, Antarctica is dominated by ice supporting vast seal and penguin populations. A rise in temperature would break up coastal ice, endangering the seals and the penguins, and the populations of krill which graze on algae attached to the sea ice.

Arctic and Antarctic organisms may also be suffering from the decrease in ozone levels within the stratosphere and the associated increase of ultraviolet radiation around the poles. The radiation is particularly harmful to marine algae and crustaceans such as shrimps, which form a fundamental part of the polar food chains.

The ozone hole

The stratosphere has a layer of ozone gas shielding the Earth from ultraviolet (UV) rays from the Sun, which can wither plants and cause skin cancer in humans. Ozone is a natural gas made when sunlight makes three oxygen atoms combine. It is destroyed by the chlorine in CFCs (chloroflourocarbons). Rising levels of CFCs in the air have made the ozone layer perilously thin, and a "hole" appears for a few months over Antarctica every spring. Polar wildlife could soon suffer badly.

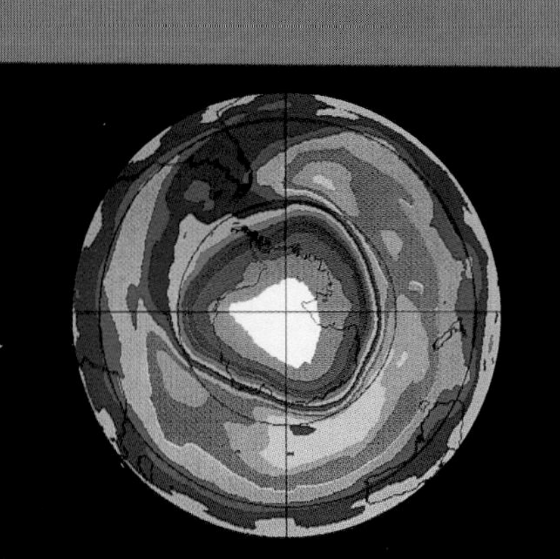

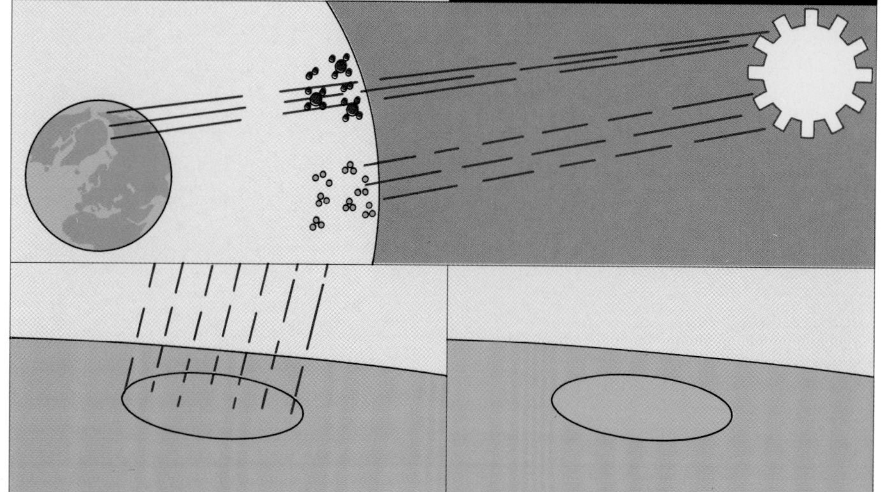

Sunlight splits chlorine off from CFCs (red dots). With chlorine present, ozone molecules break down into oxygen faster than they combine, and so ozone levels drop. Phytoplankton in the sea are thus exposed to lethal UV radiation.

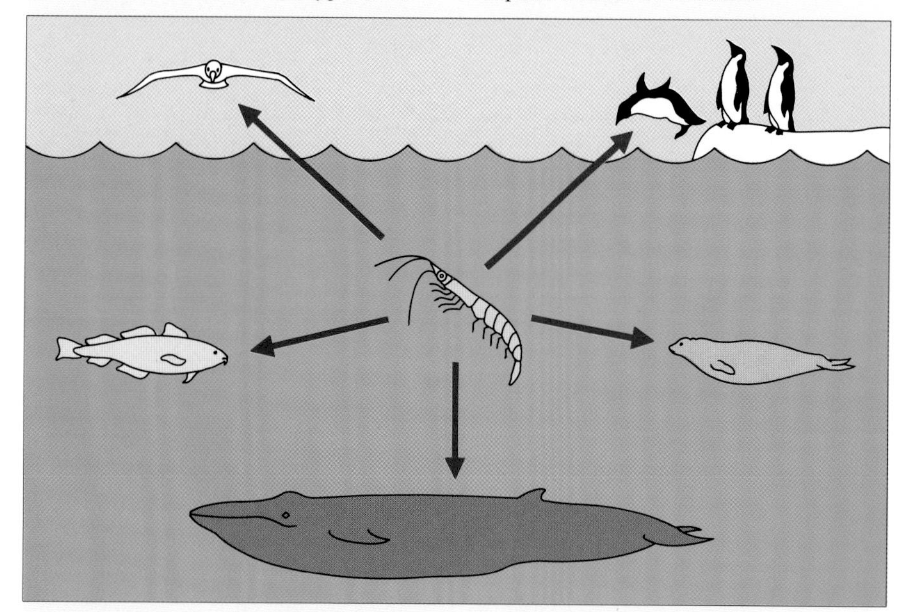

Phytoplankton are the main food of shrimplike crustaceans called krill. Krill, in turn, are at the center of the Antarctic seas' food web, forming the staple diet of many creatures including blue whales, fur seals, albatrosses, and penguins. If the thinning of the ozone layer damages phytoplankton, it will have profound repercussions right the way up through the food web.

LIGHT PENETRATION

When light enters the sea, all wavelengths do not penetrate to the same depth and below about 650 feet (the aphotic zone) virtually no light penetrates at all. Attenuation or reduction in the light intensity with depth is due to both scattering and absorption of the light, with red light penetrating the least and blue light most. In shallow coastal waters containing high levels of suspended material, light penetration is considerably less than in clear oceanic waters. In the aphotic zone, the only light comes from the bioluminescence of the creatures living there.

Light in water

165 ft

330 ft

490 ft

650 ft

6 | 7 | 10

0-3,300 feet deep

6 | 7 | 8 | 9 | 10

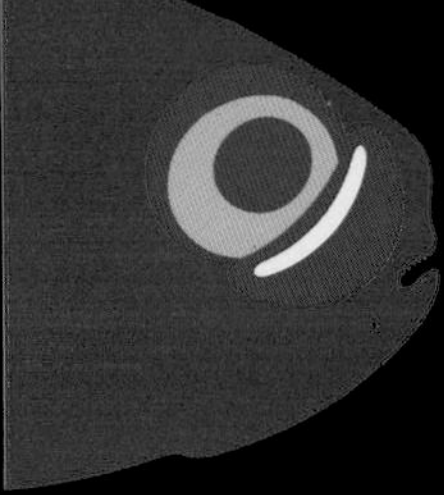

Flashlight fish

COLOR SCHEMES

Since different light wavelengths penetrate to different depths, the color of animals can appear to be quite different according to whether they are in or out of the water, or depending on the depth at which they are swimming. A red fish appears red at the surface because its pigments absorb all the other wavelengths that make up white light and reflect only the red wavelengths. When a red-colored fish is below about 60 feet, since no red light is present to be reflected by the animal's pigments, it appears to be black. Many small crustaceans that live in surface waters are transparent, whereas related deep-water species (9E) are orange or red because in the regions they inhabit they appear black, and so avoid detection by predators. In sunlit surface waters most fish are silvery, often with the upper body darker than the lower. This pattern of countershading provides protection from predators above and below, since when they are seen from below the silvery lower body is camouflaged against the brightly lit surface of the water. The darker upper body makes fish more difficult to detect by birds hunting above the surface.

The size of predators' eyes increases with depth to accommodate poor visibility, but in the aphotic areas eyes are of little use, and prey can be detected by chemical signals, by small electrical currents generated when muscles contract or by currents created by swimming movements. Other species use a sonar system to communicate or locate their prey. Chemical signals are also important for finding a mate, while some species like the flashlight fish (6E) are bioluminescent and use their light source as a signal to attract potential mates. They can switch their light on and off by sliding back and forth a curtain of skin across the luminous organ. Bioluminescence may be produced chemically by the fish itself, or by symbiotic bacteria, as in the flashlight fish.

Light-producing organs may also be used as lures for prey, a technique that has been perfected by the angler fish (9B), and by the black sea dragon (12E).

0–300 feet deep

Ecosystem Profile *The Southeast Pacific*

The ocean region offshore from Chile and Peru has a narrow continental shelf, and its waters are dominated by the Peruvian Current, which flows from origins in the Antarctic Circumpolar Current northward along the Pacific coast of Latin America. The influence of trade winds causes the surface water to flow away from the land at a point opposite the northern coast of Chile, allowing deep, cold, rich waters to rise to the surface.

Microscopic phytoplankton

SURFACE WATERS

This area supports one of the most productive fisheries in the world with an estimated sustainable yield of 12.6 million tons a year. The high productivity results from the upwelling of cold nutrient-rich water brought to the surface. The primary producers of the ocean, the microscopic phytoplankton, dinoflagellates, and diatoms use sunlight to photosynthesize and produce complex organic molecules that form the basic food for all higher organisms. Photosynthesis can only occur in the presence of sunlight – in the upper part of the water, known as the euphotic zone. For most of the world's oceans, the limits to primary productivity are the nutrients, nitrogen and phosphorus, which occur in high concentration at depths below the level of light penetration, where phytoplankton cannot make use of them. So areas of upwelling, where nutrients are brought to the surface, are particularly important in the open ocean, and in these regions the primary productivity is high. The phytoplankton in such regions also tend to be larger than those characteristic of the low-nutrient areas of the open ocean. Many species are colonial and achieve a large enough size to be fed on directly by some small fish, as well as larger zooplankton. In open ocean areas the process is more convoluted: the minute phytoplankton are eaten by tiny zooplankton, which in turn feed slightly larger predatory zooplankton before finally being consumed by zooplankton large enough to feed fish.

The efficiency of food chains in upwelling areas, therefore, is high and small fish, such as anchovetta (5B) and sardine, which feed directly on phytoplankton and herbivorous zooplankton, can themselves be harvested directly for human consumption or for the production of oil and fishmeal. Sea birds, such as gulls (4A), the guano (fertilizer) producing cormorant, the booby, and the brown pelican (1A) also feed on these fish. In 1950 the population of these birds was estimated to number 30 million. The smaller fish also form the food of larger predatory fish species, which in turn are eaten by dolphins (2C), sharks (4B), and toothed whales such as the killer whale and the sei whale (1C).

OCEANS

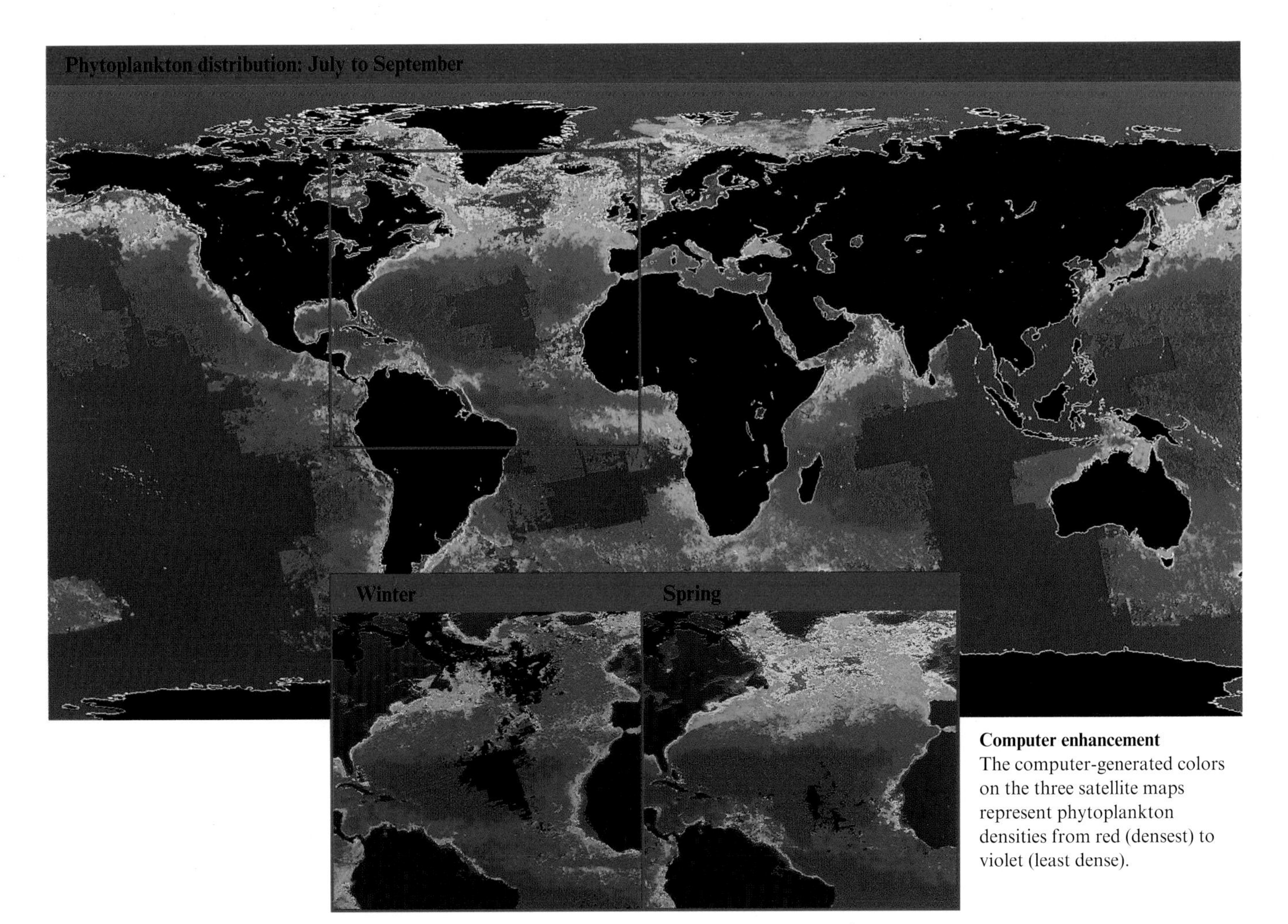

Computer enhancement
The computer-generated colors on the three satellite maps represent phytoplankton densities from red (densest) to violet (least dense).

Despite appearances to the contrary, the oceans do not constitute one uniform environment, but vary considerably in both their chemical and physical characteristics, and with depth and geographical latitude.

Light, pressure, temperature, and the availability of nutrients all vary with depth, affecting the distribution and abundance of oceanic organisms. In relation to latitude, species diversity gradually declines with distance away from the equator (as it does on land): the Indo-West Pacific is the most biologically diverse marine area in the world. Surface temperature is also greatly affected by latitude – and because the seas are in constant motion, both heat and salt are moved around the ocean basins. The natural swell causes mixing at the surface, and heat absorbed there tends to sink downward, with the result that the overall range of temperatures in ocean environments is very much smaller than that in terrestrial environments.

Perhaps the most obvious variation regulated by depth is the attenuation of light, which limits the zone in which photosynthesis is possible to a stratum close to the surface. Below 300 feet, even in the clearest water, only blue light penetrates; and below about 700 feet, almost no light penetrates at all.

Yet primary production is not uniform in all the surface waters of the oceans. Much of the open ocean, for example, is the marine equivalent of a desert because of the restricted presence of nutrients at the surface. The nutrients nitrogen and phosphorus may be brought to the surface in zones of deep-water upwelling, and as a result productivity is high in such areas, which support the most highly productive commercial fisheries.

The presence or absence of nutrients is of critical importance to oceanic production in surface waters. Coastal and near-shore areas that receive deposits of nutrients from the land thus tend also to be quite productive: it has been estimated that some 75 percent of the total world commercial fish catch comes from within 5½ miles of the shore. However, not all coastal waters receive equal amounts of nutrients. They may vary according to the levels of local rainfall, the volumes of run-off from the land, and the nature of the soil and underlying geology.

Also highly significant to phytoplankton production is the water temperature. Zones of productivity for this reason shift with the seasons, reflecting both the temperature at the water surface and the angle – and thus the intensity – of the sunlight. In temperate regions, the spring bloom of phytoplankton is followed by a decline in productivity as nutrients are removed from the water. Once production stops, or is severely restricted during the winter season, the rate of nutrient input and release into the water becomes greater than the rate of uptake by the phytoplankton. Phytoplankton production increases as the temperature rises

Most of the ocean is a liquid desert, with life forms few and far between (left), but there are oases of life such as the Sargasso Sea (top) or the deep sea hydrothermal vents (above).

again in spring, and the bloom occurs once more, feeding on the now available nutrients. Water temperature also plays an important role in determining the distribution of several types of marine and coastal ecosystem – such as coral reefs and mangroves.

The salt in the sea

Salinity also varies considerably according to geographical location. Areas of high surface evaporation and low rainfall are characterized by higher salinity than areas of low surface evaporation and high rainfall. The surface waters of the Bay of Bengal, for example, which are influenced by the massive water input of the combined Ganges-Brahmaputra (Meghna) River, contain around 34 parts per thousand (ppt) of salt, compared with 36 ppt in the surface waters of the Arabian Gulf.

The Caribbean Sea is an area of high salinity, despite the inputs of the Orinoco and Mississippi Rivers, because the evaporation is so fierce. The oceanic area off Southeastern Asia, meanwhile, is characterized by low salinities of less than 33 ppt due to the enormous river inputs in this region. Such variations in salinity – and in freshwater inputs – also affect the distribution of animals and plants throughout the ocean basins.

Salinity, density, and temperature of the water are all related, but these factors differ considerably across the surface of the oceans. As water cools, its density increases and it tends to sink – which is what happens to the high-salinity water of the Gulf Stream as it reaches the edge of the Arctic Ocean. This cold, dense water then flows south below the surface. Different water masses are discernible at different depths in the ocean by their physicochemical characteristics and their direction and speed of flow, and each such mass of water supports a recognizably different community of oceanic organisms.

Of considerable biological interest are the eddies, or rings, that characterize many boundary currents. The Gulf Stream on its passage north, for example, meanders, and the meanders may then become cut off from the main flow of the current – as warm core eddies to the north of the main flow, or as cold core eddies to the south. The community of organisms inside each ring is characteristically quite different from the community outside the ring.

Ocean patchiness

Some areas or zones of the ocean system are utterly distinct from those surrounding them. One such is the Sargasso Sea, an area of relatively little water movement, characterized by the floating *Sargassum* weed and its associated community of fascinating animals camouflaged to resemble the algae among which they hide.

It is not only the surface waters that are composed of a mosaic of different habitats and communities. The deep ocean is now known to support its own areas of high productivity, surrounded by less productive systems with less diversity. One such deep-sea system occurs around the deepwater ocean vents, where underwater volcanoes discharge superheated water and lava out on to the ocean floor. Such vents are quite widely distributed along the fracture and fault lines of the mid ocean ridges, where they support unique communities of organisms dependent on chemotrophic bacteria. Instead of fixing sunlight by photosynthesis, these bacteria derive their energy from sulfur sources and form the basis of a complex food chain involving a community of filter-feeding mollusks, worms, and echinoderms.

Seamounts often support equally unusual communities of animals. The mounts may be tall enough to reach close to the surface and form isolated areas of high productivity in the surrounding ocean. The orange roughie fishery of New Zealand has unfortunately all but wiped out the populations of fish on some more local seamounts through unsustainably, overrapid commercial exploitation within areas that were already small and vulnerable.

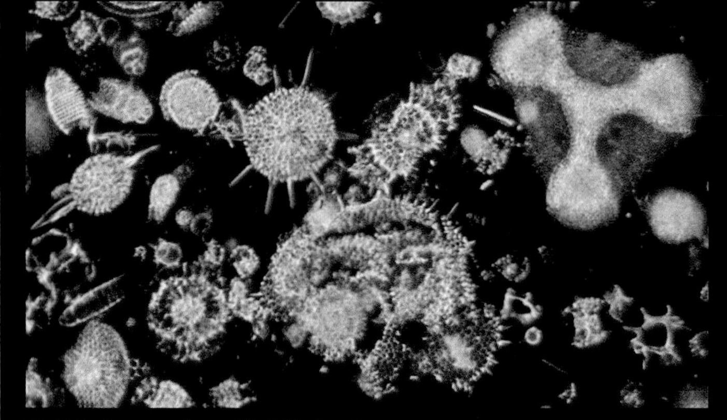
Radiolarian ooze (magnified 80 times)

THE SEA FLOOR

The bottom of the ocean, or benthic zone, is populated by sessile animals attached to the ocean floor, creeping animals that wander in search of sustenance, and burrowing animals that live under the soft sediment of the ocean bed. Just like the animals in the pelagic zone, the benthic community is also widely distributed, low in density, and high in species diversity. Taking the pelagic and benthic zones together, they contain over 10 times the number of species to be found in the euphotic zone – an estimated 2,000 species, as opposed to 200.

For their sustenance these deep-sea creatures depend on the constant fall of "fecal rain:" the decomposing bodies of dead animals and other waste material that sinks from the lit surface waters. In contrast to the continental shelves (the drowned fringes of continental land masses), which are covered by sediment from land erosion, the sediment of deep ocean basins is composed of skeletal material derived from the planktonic community of the ocean surface.

Radiolarians, diatoms, and foraminifers form characteristic oozes in various areas; in deeper basins there are red clays that represent the finest particles derived from land erosion. The rain of organic material forms the concentrated energy source that fuels this community. Those animals that do not depend on this fecal rain are the groups that surround ocean thermal vents, which occur along fault lines characteristic of mid-ocean floor ridges.

In these areas molten lava and superheated water is expelled from the mantle below; the metals in solution precipitate on contact with colder water to form deposits of metalliferous muds. The materials frequently have a high sulfur content and specialized bacteria use this as a source of chemical energy. The bacteria in turn are eaten by a range of filter-feeding worms and mollusks and a complex community of species develops.

For most creatures that live in the benthic, however, the sinking detritus from the euphotic zone is the primary energy supply, and species that feed directly on it include sea pens (17G), sea lilies, and glass sponges, who feed by filtering the material in suspension; others, like the brittle stars and sea cucumbers, pick particles of organic material from the surface of the sediment or burrow into the sediments, ingesting them indiscriminately and extracting any organic material as they pass through.

Filter feeders decline in abundance with depth, but the burrowers are found in even the deepest ocean basins. There the soft-bottom-dwelling community includes a wide variety of species of fish and crustaceans which have extremely long appendages to support themselves on the sediment surface.

The unusual *Bathyterus* or tripod fish (18E) has, as its name suggests, three elongated "legs" which it uses to "walk" on soft sediments. As a consequence of the low food availability on the bottom of the ocean, the growth rates of deep sea animals are very low. Some bivalve mollusks from depths of 10,000 feet off the North American coast are only an inch or so long even though they are at least 250 years old.

16 17 18 19 20

20,000-33,000 feet deep

E F G H

16 18 19 20

LIFE IN THE DEPTHS

There are considerable zones of water below the euphotic area and above the ocean floor, which form the deep pelagic regions of the world's oceans. Although oceanographers disagree on the precise division of this region, three areas are generally recognized: the mesopelagic occurs between 650 feet and 6,500 feet; the bathypelagic is between 6,500 feet and 19,500 feet; and the abyssopelagic is found below 19,500 feet. The aphotic pelagic zone (below 650 feet) is a relatively stable environment, with only a slight variation in temperature and salinity throughout all the world's oceans. The temperatures are also very stable, but low (between 35° and 37° F) in the abyssopelagic zone.

A major problem for creatures living at these depths is the scarcity of food, since the entire pelagic community is dependent on organic material produced in the euphotic zone above. (Exceptions are the unique communities surrounding ocean floor thermal vents.) Food supplies decrease with depth, and the density of species also drops, but the diversity is quite high. Many deep-water pelagic species have a wide geographic distribution and occur in all oceans. Predators in this region, such as hatchetfish (10E) have low metabolic rates due to the low temperatures and infrequent meals. In order to conserve energy they tend to have reduced muscle mass and long, thin bodies, like the cookiecutter shark (10B), and the viper fish (7C). In contrast, their heads and jaws are often disproportionately large and the body is very flexible. This allows fish, such as the *Chiasmodon niger* (15F) and the gulper eels (15D), to swallow large fish, a useful skill in these sparsely populated waters. The head of the appropriately named gulper eel, with its massively extended jaws, dwarfs its whiplike body. *Chiasmodon* goes even further: it can partially dislocate its jaws, and its stomach distends to allow it to swallow prey as large as itself.

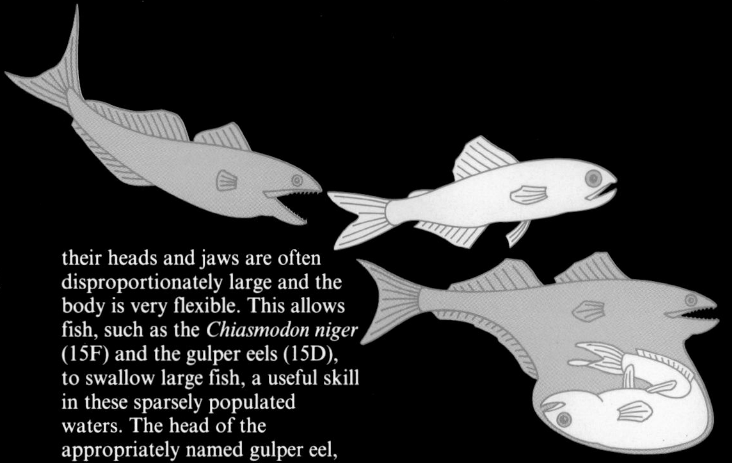

C. niger **devouring pr**

11 12 13 14 15

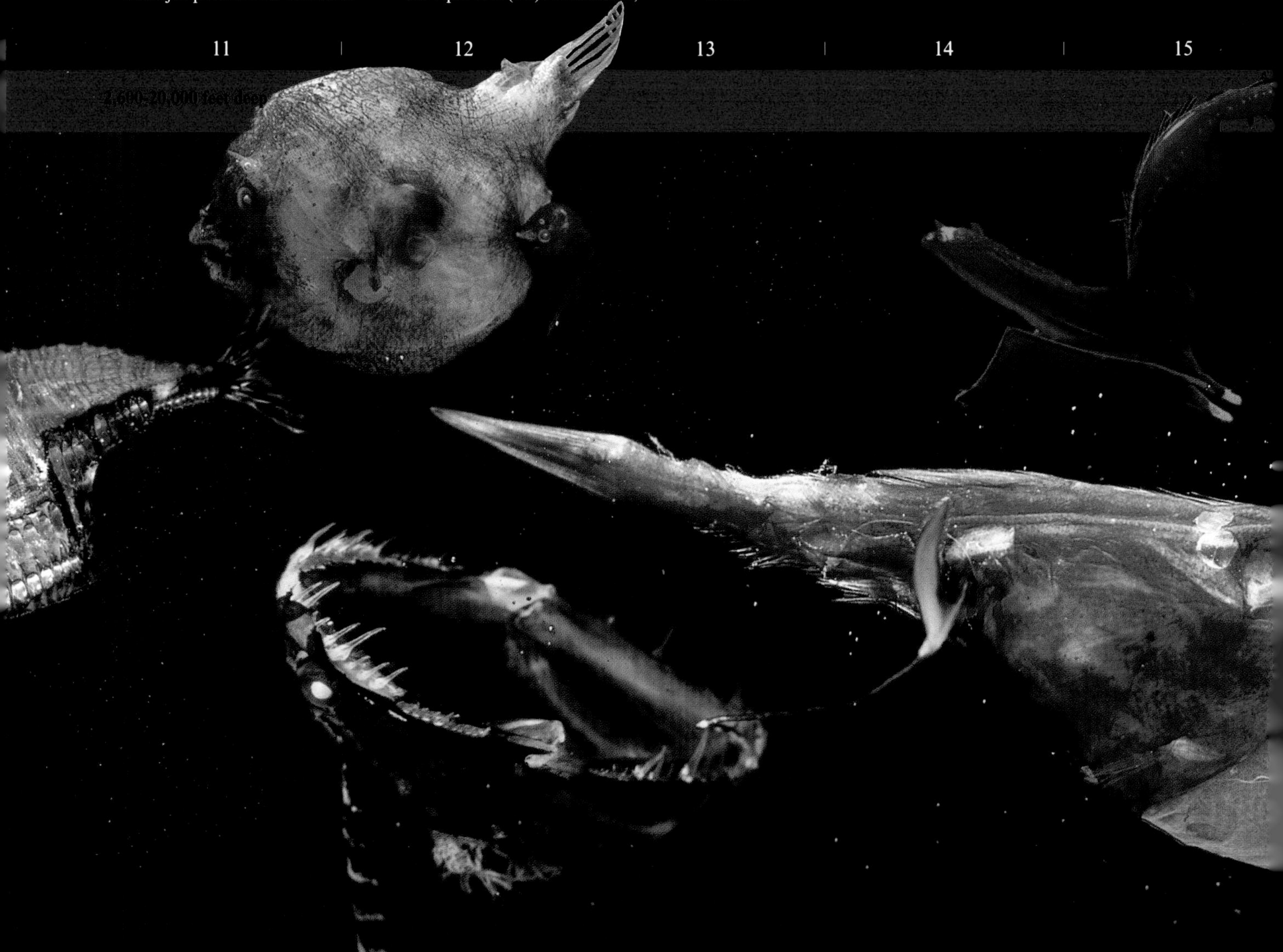

2,600–20,000 feet deep

11 12 13 14 15

UPWARD MOBILITY

In addition to the regular inhabitants of the deep, sperm whales are known to descend to depths of as much as 6,500 feet in search of the giant squid on which they prey. One exploratory deep water submersible even encountered a swordfish, indicating that a surprising number of surface dwelling species may penetrate the depths in search of food.

Schools of hatchetfish (9C), lanternfish and others, as well as squid and small crustaceans (9E), come close to the surface at night and return to deep water by day, thus having to cope with huge pressure variations (from one atmosphere at the surface to over 30 atmospheres at 1,000 feet).

Various species of anglerfish spawn at great depth, but their eggs float to the surface before they hatch and the larvae feed themselves on plankton before sinking back to around 3,300feet. There, several species of anglerfish have their own solution to the problem of finding a mate in the far-flung pelagic. The tiny male attaches himself to the body of the larger female (12D), and spends the rest of his life carried about by her and nourished by her bloodstream.

OCEAN DYNAMICS

The key to survival

- Water is transparent, so where there is not too much sediment, light can penetrate for photosynthesis
- Oxygen is available at all depths
- The constant exchange of energy between the deep, cold layers and the warmer surface

Forces for change

- Ocean basin-scale circulations which affect nutrient upwellings
- Water temperature and light fluctuations affecting seasonal production
- Pollution
- Fishing and overfishing affect the distribution and survival of species, and some techniques physically destroy habitats
- Nutrient enrichment causing algal blooms and red tides

There is so much water in the world's oceans that they seem almost immune to change. Even coastal waters lag far behind the seasons, slow to cool in winter and slow to warm in summer. Out in the open ocean conditions hardly vary from month to month, let alone from year to year. Down below 660 feet in the dark mesopelagic zone and the even darker bathypelagic zone toward the ocean bed, conditions almost never change. There is no difference here between night and day or winter and summer, and the water is almost always at 36°F. Yet the ocean is a dynamic ecosystem like any other, and despite its titanic size, it can and does suffer from shocks and impacts, natural and human.

There are actually regular seasonal variations in the abundance of life reflecting the changing availability of nutrients in surface waters where most marine life is concentrated. In certain places in the oceans – usually near the western coast of continents (especially the Americas) – nutrients are continually swept up to the surface by temperature differences in the water. The nutrient-rich water in these regions stimulates prolific phytoplankton growth, so there is a proliferation of all kinds of marine life, as fishermen are aware. Most of the small pelagic fish – sardines, anchovies, and herring-like fish – come from such areas and form around 20 percent of the total world catch.

Upwellings off the Americas coast tend to continue all year round. But in some parts of the world, upwelling is seasonal. The coming of the monsoon to the lands around the Indian Ocean, for example, coincides with the appearance of an upwelling off the coast of Somalia and Oman, which encourages a sudden proliferation of plankton and fish, and also allows unique communities of cold water macroalgae and abalone to live on the rocky shores of Oman.

Major oscillations

Just as ocean ecosystems can respond to seasonal variations, so can they respond to longer-term, or more sudden and dramatic events, such as the coming of the cold Pacific ocean current El Niño every ten years or so. Although it is now possible to predict on a shortterm basis the onset of El Niño, noone yet knows what it is that makes El Niño appear. Nevertheless it seems clear that such dramatic shifts in ocean currents can have important implications for marine communities and ecosystems, particularly in areas of upwelling and along continental shelves.

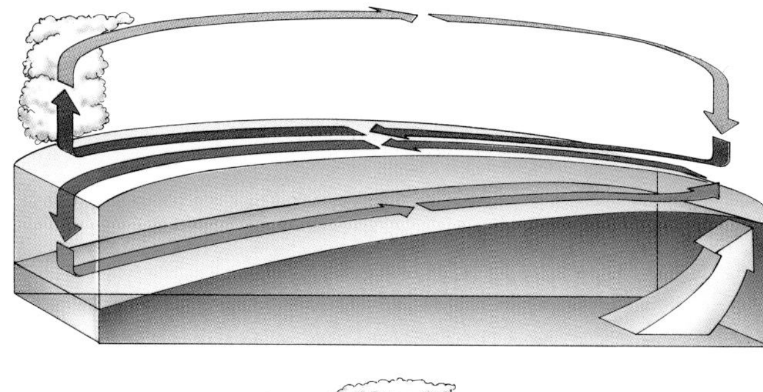

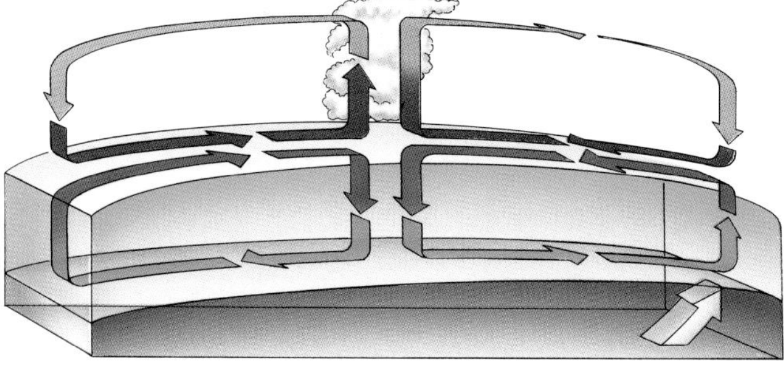

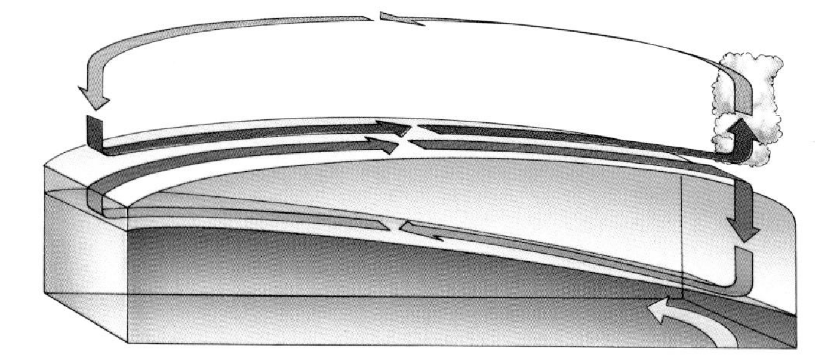

El Niño

The surface waters of the Pacific Ocean at the latitude of northern Chile and Peru normally flow in a westerly direction, increasing in temperature as they cross the open ocean and forming a pool of warm water in the western Pacific. However, during El Niño (The Child, so-called because it usually happens around Christmas) events, wind patterns weaken and current circulation changes, shutting off the cold-water upwelling and therefore the supply of nutrients rising to the surface. The results of these events are dramatic: not only can the sea level rise 20 inches along the coast and drop 5.5 inches in the western Pacific, but primary production crashes, fishery yields decline, and bird population dependent on the anchovetta and other small fish starve and fail to breed. During the 1972 and 1982/83 El Niño the fish catch dropped to one sixth and one six-hundredth of the normal landings respectively. The Peruvian catch of 13 million tons in 1970 dropped to 1.2 million during the 1972 El Niño, and fish living close to the sea floor, such as *Sciaena deliciosa*, shifted from their normal range of 7.5 miles off shore to some 100 miles due to the increased water temperature. Other species, like the Peruvian scallop, become more abundant because spawning and larval survival are higher in warmer waters. During the 1982/83 El Niño, catches of this scallop increased to 10,000 tons from the normal level of 1,000 tons. Populations of cormorant, booby, and brown pelican, which numbered in excess of 30 million during the 1950s, declined to an estimated 6 million after the 1972 phenomenon and dropped even further to an estimated 300,000 after 1982/83.

- Air current in the atmosphere
- Air current along the surface of the sea
- Water movement caused by air currents
- Water movement along warm water/cold water boundary
- Deep-sea cold-water upwelling

Human impacts

Humans have long used the oceans as a garbage-dump, assuming that everything will be diluted in their vast waters, and every day huge quantities of effluents and waste are pumped into the sea. But their effects can often be far-reaching. All kinds of toxic chemicals are regularly washed into the sea including cadmium, lead, pesticides, and herbicides. It is thought that mercury pollution in particular can inhibit the growth of all kinds of plankton, especially phytoplankton, the key species right at the beginning of the food web. Pollution in the North Atlantic has reduced the number of species of plankton, and if plankton are affected, then pollution could damage life up through the food web. Chemical pollutants have also been implicated in the death of seals in the North Sea. No-one knows what effect the huge quantities of high-level radioactive waste once dumped at sea will have. Low-level waste is still being dumped.

Even "natural" waste such as sewage may have a damaging effect. Sewage discharged from Californian cities has contaminated over 1,400 sq miles of the oceanbed, degrading shellfish and crustaceans, killing beds of kelp, and causing disease in fish. Changes in phytoplankton species, too, appear to happen more frequently in coastal waters because of increased levels of nutrients coming from sewage effluence. This causes the unsightly green scum near the surface found off tourist beaches in the Adriatic. Increased nutrient discharge is also blamed for the rising frequency of toxic algal blooms and red tides in many areas around the world.

Overfishing, too, has had a dramatic impact on ocean ecosystems. Even once-common fish like cod are now quite rare, while modern trawling damages marine beds, and tuna-fishing methods often kill dolphins and other sea mammals. The sad fate of whales is all too well known.

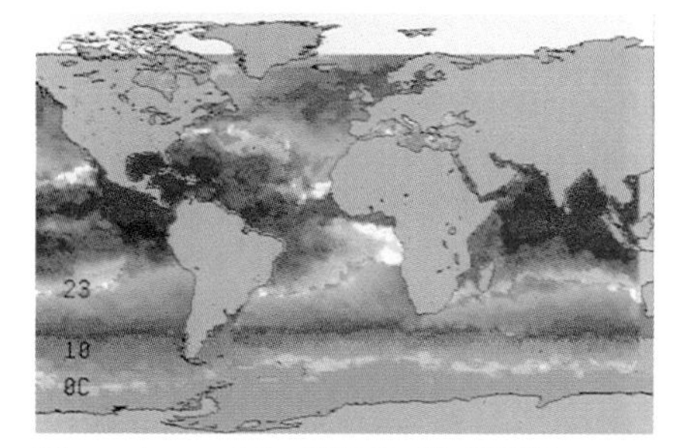

Thermal map of the 1982 Pacific El Niño event

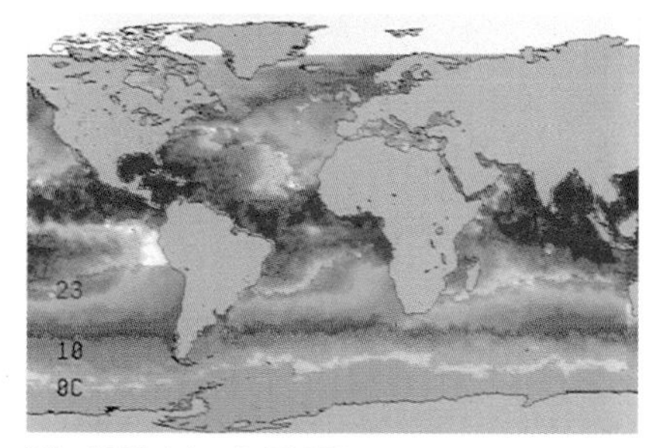

The 1983 Atlantic El Niño event

A small, sunken world

Just as deep sea thermal vents are centres of high density and diversity on the sparse ocean floor, the dead bodies of large organisms such as whales suddenly provide a huge input of energy into a system that normally depends on smaller particles; they become, in effect, selfcontained worlds, supporting large populations of bacteria that decompose the animal remains and are a food source for many filter-feeding organisms.

The marine iguanas of the Galapagos are among the many species decimated by Pacific El Niño events

TIDE POOLS

The high intertidal zone is considerably more diverse than the splash zone. Permanent tide pools are often found and serve as refuges for a wide variety of crabs and mollusks during periods of low tide. One species of hermit crab, *Pagurus samuelis* (3B), is a common resident of these tide pools and uses the turban shell for its home. The little rock crab, *Pachygrapsus crassipes* (4C) is an active scavenger over the rocks at night and hides in crevices and pools during the day. Here also the *Balanus* barnacle (3D) replaces the *Chthamalus* of the splash zone.

Seaweeds also appear in this region, which is dominated by the blue-green algae and green species. The brown rock weed *Pelvetia* occurs around the middle of the zone and serves as a nursery area for the juvenile *Littorina scutulata*. Other mollusks on the rock surfaces are the limpets, with *Acmaea persona* occurring high up, gradually being replaced lower down by *Acmaea digitalis.* A limpet's shell shape seems related to its position on the shore: species that live in higher, drier habitats seem to have taller shells, perhaps to enable them to store more water.

6 | 7 | 8 | 9 | 10

Middle intertidal

6 | 9 | 10

THE SPLASH ZONE

The upper reaches of the shore are home to a small periwinkle, *Littorina planaxis* (1D), which is well adapted to resisting desiccation. Indeed it can survive extended periods out of seawater, over 40 days in the case of some individuals. These periwinkles only require brief periods of immersion in order to wet their gills and allow them to breathe, and in areas of heavy surf *L. planaxis* may be found several yards above the high tide level.

Its ability to resist the drying action of the wind depends on the presence of a horny operculum, like a manhole cover, that fits the mouth of the shell and provides an airtight seal. This ensures that little or no water is lost from the surface of the gills, and the animal emerges to feed only when the temperature is low and the humidity of the air high.

At lower levels on the shore *L. planaxis* is replaced by a close relative *Littorina scutulata*, which is slightly smaller and less resistant to desiccation. Both these animals feed by scraping the microscopic film of algae from the rock surfaces by means of a filelike radula – a ribbon of flexible tissue, covered with rows of hard, angled teeth. The radula is constantly worn away by its abrasive action, but is continually replaced from behind.

Littorina planaxis closing its operculum (green disk)

INTERZONAL FEEDERS

The starfish (11F), sea urchins (12F), and brittle stars (18G) are all related, and all possess a unique mechanism for locomotion: the water vascular system.

Along the underside of the arms of starfish, and in rows over the body of sea urchins, there are groups of small tube feet, often with a sucker disk at the free end of each. These can be extended by having fluid forced into them from a small internal reservoir. The feet are first stretched and then attached to a surface. Through a contraction of the muscles, the tube foot is shortened, pulling the animal forward. A predatory sea star will also use its arms to pry open the shells of mussels and clams, before extruding its stomach directly into the shell of the mollusk to digest its prey.

Supralittoral | Upper intertidal

1 2 3 4 5

A B C

Ecosystem Profile *The Californian Coastline*

The Pacific coast of North America is dominated by the swells of the Pacific Ocean, which, in combination with the coastal configuration, result in areas of high exposure to wave action, as well as more sheltered areas in semienclosed bays and the lee of headlands. Some stretches of coast are dominated by rocky conditions such as those shown here, while others are characterized by sand and mudflats and eelgrass beds.

The coastal zones and barnacle distribution

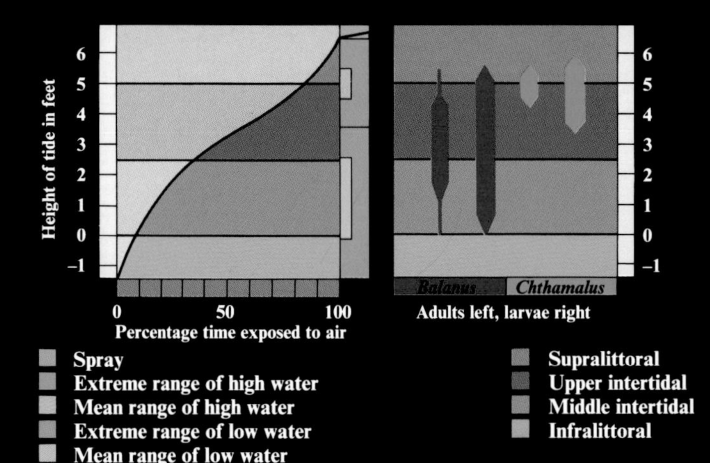

ZONATION

The Californian coast can be divided into recognizable zones, each of which is characterized by a particular community of organisms. The splash zone, supralittoral or *Littorina* zone occurs at the head of the shore, where the area is only inundated by extreme high tides and is mostly subjected just to spray and wave splash. In this area only the hardiest of marine organisms, such as the pill bugs, are found. The upper intertidal appears between the mean high water level and the mean sea level. It is dominated by barnacles and other animals capable of withstanding long periods of drying by the wind and heating by the sun. The middle intertidal extends from the mean sea level to the mean low water level and is characterized by animals that require a regular, twice daily covering by the tides. Below this is the infralittoral fringe or low intertidal that extends from the level of mean low water to extreme low water. This zone is dominated by the kelps, *Laminaria* (12A), and is home to animals that cannot withstand exposure for any length of time.

This pattern of zonation, although widespread and generally applicable, varies according to a number of factors. In exposed shorelines the zones tend to be wider and higher than in sheltered locations, where the entire set of zones is lower on the shore and the splash zone may be regularly and frequently inundated by the highest tides.

Zonation is not only affected by exposure but may be altered by other factors such as competition and predation. Two barnacles, *Balanus balanoides* and *Chthamalus stellatus,* are found together on rocky shores, with the former dominating the rock surfaces of the lower and middle intertidal and the latter occurring as adults only in the splash zone. The larvae of both species settle over a much wider range of zones, but in the splash zone the *Balanus* die due to desiccation, and in the upper intertidal the *Chthamalus* are eliminated by competition for space with the faster growing, lower profile (and thus harder to dislodge by wave action) *Balanus.* The lower limit of the *Balanus* is controlled by predation from a drilling mollusk *Thais*, which cannot resist exposure for too long so its distribution is confined to the lower infralittoral zone.

COASTLINES

Coastlines are just the thin strip where the land ends and the sea begins, a narrow buffer zone between the marine world of the ocean and the terrestrial world of the land. Yet they include some of the most productive of all biological communities, alive with seabirds swooping through the air, shellfish and crustaceans burrowing in the sand, fish darting through the water, and much more beside. Coastal land, and the shallow coastal waters of the continental shelf occupy barely 8 percent of the surface, but account for 26 percent of all biological productivity.

Yet despite its richness, the coast can be a very exacting, tough environment. The shoreline in particular is constantly battered by waves, soaked in salty spray, inundated twice daily by the rising tide and then desiccated by the Sun. To survive this harsh regime, shoreline organisms have had to develop a whole variety of different adaptations.

Coasts are enormously varied environments. A sheer cliff face can give way to a sandy beach or a mudflat and then to cliffs again within the space of a few hundred yards. They are changing all the time, too, not just from year to year as cliffs are undercut and beaches are reshaped by crashing waves, but from hour to hour as tides ebb and flow, and from moment to moment as waves roll up and down the shore. Ecologists tend to divide coasts into rocky, sandy, and muddy, but these groupings hide an almost infinite variety.

The constant shifting of the boundary between land and sea created by the tides has a profound effect on the shoreline ecosystem. Tides are created when the gravitational pull of the Moon distorts the Earth and its parcel of water slightly, stretching them into an oval. This creates a tidal bulge over the part of the Earth nearest to the Moon, and a second "bulge" (in reality a tail, or wake) on the side of the Earth farthest from the Moon. Tides rise and fall twice daily as the Earth's rotation swings each part of the ocean into line with the Moon, then swings it away again. Twice a month, the Sun lines up with the Moon and their combined pull creates very high and low "spring" tides. In between come the minimal "neap" tides when the Sun is at right angles to the Moon.

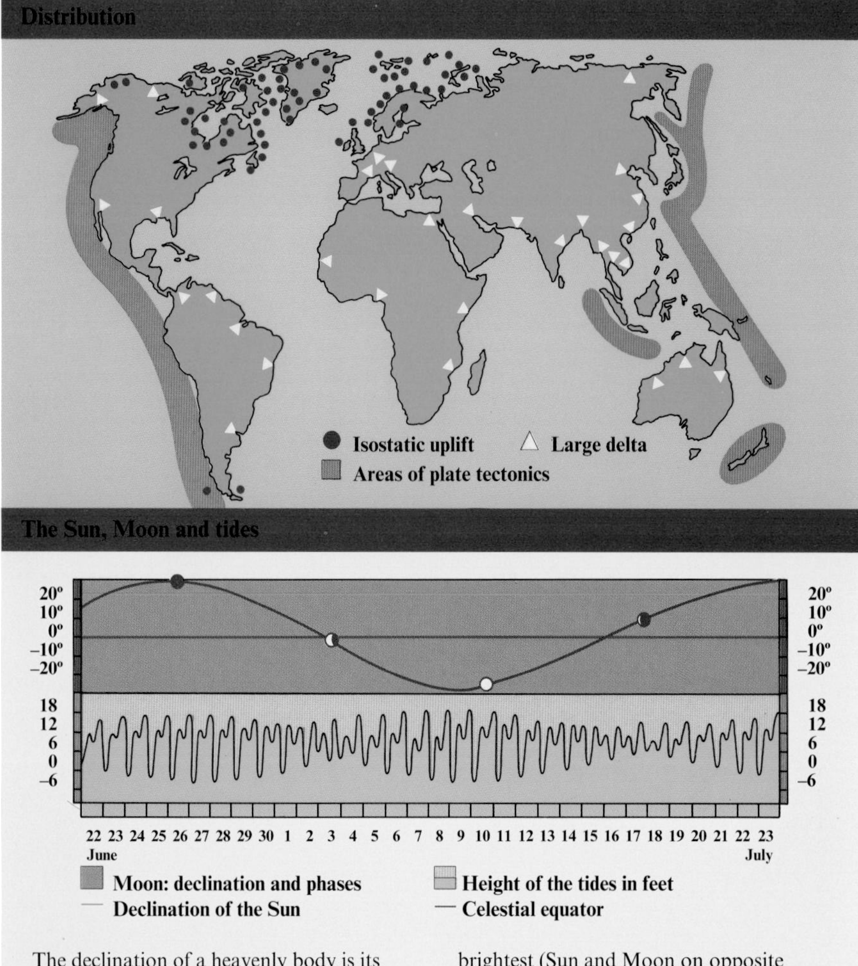

The declination of a heavenly body is its angle from the celestial equator (the plane of the Earth's equator projected out into space). The strongest tides, the spring tides, occur when the Sun and Moon are aligned. This in turn corresponds to the those periods when the Moon is at its brightest (Sun and Moon on opposite sides of the Earth), and at its darkest (Sun and Moon on the same side of the Earth). There is actually a slight delay between these periods and the corresponding high tides, as the tidal bulge of water lags behind the orbiting Moon.

Shoreline zones

On some coasts, the tidal range is over 50 feet. On others, tides are imperceptible. But wherever there are marked tides, they create a distinct zonation of plant and animal communities up and down the shore. On rocky temperate shores, this can be so pronounced that the shoreline appears to be striped.

High up rocky shores, beyond the reach of the highest tides, is the splash or supralittoral zone. Here organisms can tolerate frequent soaking with salt water, but are essentially terrestrial. Below the supralittoral is the intertidal zone, covered and uncovered twice daily by the tides. Lower down still is the infralittoral fringe, uncovered only by the lowest spring tides. Beyond the infralittoral is the sublittoral zone, the open sea.

Although tides are the main control over the distribution of life on the shore, there are other factors at work. The degree of exposure to crashing waves is important as well. On sheltered shores in the North Atlantic, for example, a kind of seaweed called the knotted wrack thrives; but on wave-pounded shores, the knotted wrack is replaced by the more robust bladder wrack.

The character of the seashore – rocky, sandy, or muddy – is important too. Each type of seashore has its own characteristic species, as do different types of rock, sand, and mud shore. Granite shores, for example, have

The rise and fall of the tides, and the pounding of the waves, shape the rock formations of the coastline as well as the lives of the inhabitants

crevices ideal for mollusks, while limpets like the smoothness of some basalt rocks. Zonation is less obvious on muddy and sandy shores, which can appear barren because most of the inhabitants burrow at low tide. This means that they have fewer problems with desiccation than rocky-shore dwellers, but must be able to respire in conditions where oxygen is not readily available. The size of sand grains influences the spread of burrowing animals because tunnels tend to collapse in coarse sand. It also influences the distribution of animals such as the fiddler crab which feeds by picking up grains and drawing off the microscopic coating of algae through the setae (hairs) around its mouth – but only certain size grains fit the setae.

Tidal clocks

Sea tides not only influence whereabouts on the shore each species of organism finds its home; they also influence their life cycles. So dramatic are some of the changes wrought by the tides on the shoreline environment that many organisms have learned to adapt their biological rhythms to match the ebb and flow of the tides. When the tide comes in, the parched, vulnerable life of the shore is bathed in refreshing water, revitalized by its store of food, oxygen, and salt, and hidden from aerial predators. But it is also exposed to the battering of the waves, and the attentions of marine predators.

Many shore crabs, like the European shore crab, live by a tidal clock. As the tide ebbs leaving the beach exposed, these crabs hide among the seaweed or in rock crevices to avoid being spotted by predatory birds such as the crab plover. But as the tide flows in again, they emerge to feed, safe beneath the water. Some crabs emerge to feed at low tide when this occurs at night. Fiddler crabs feed at low tide all the time, despite the risk, because they perform their mating display on the exposed beach.

What is remarkable about all these crabs is their sensitivity to the state of the tide. Some zoologists think that since the tides are determined by the phases of the Moon, the crabs must have an internal lunar clock, responding to changes in the intensity of moonlight.

Just as the crab's daily rhythms are dictated by the need to avoid exposure to predators at low tide, so the feeding habits of many shoreline predators are regulated by the tides. An incoming tide brings with it a range of marine animals that come into the intertidal zone to feed on its creatures and the seaweed. In its turn, low tide brings out a different set of predators; and flocks of sea birds probe the mud and soft sand to feast on burrowing worms and mollusks.

The reproductive cycles of many sea creatures are regulated by the tides, and even some deepsea fish only spawn at certain tidal states. But it is thought that they are responding to changes in the pressure, temperature, and salinity of the seawater created by the water movement rather than directly to moonlight.

One of the most striking adjustments to the rhythm of the tides is that of the Californian grunion, a small fish that lives in the Pacific off the west coast of the United States. The grunion spawns in huge numbers on sandy beaches on warm summer nights during a full or new moon. These little fish take advantage of the high spring tide to lay their eggs in the sand at the very limit of the tides. The eggs then have two weeks to hatch before the next spring tide reaches the same point on the beach and washes them out to sea again.

HIDING PLACES

The scale worms *Arctonoe* occur commensally (dependently, but not parasitically) with a wide range of animals on the shore: some species live on chitons (20G), others on sea cucumbers, and one, *Arctonoe vittata*, occurs in the mantle cavity of the keyhole limpet, *Diadora aspera* (11F). This limpet may reach lengths of 2-3 inches and the scale worm may be found curled round the foot, just inside the lip of the shell. The worm derives undoubted protection from potential predators, but other animals such as the decorator crabs rely on camouflage to avoid detection. Of the decorator crabs, *Loxorhynchus crispatus* (14E) occurs low down in the intertidal. These animals collect appropriate materials, such as sponges, bryozoa, algae, and seaweeds, which they hold in place over their backs using the upcurved fourth and fifth pairs of their legs. When in a new environment, they will discard their old camouflage and collect materials appropriate to the new location. Not only does this provide the animal with protection, but probably also enables it to stalk its prey more easily.

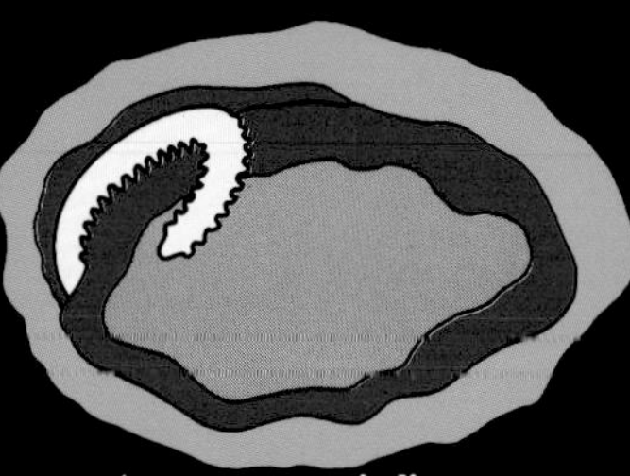

***Arctonoe* worm in limpet**

LOW EXPOSURE

In the lowest intertidal zone the diversity of the animals is high. Exposed rock faces are the home of purple sea stars, mussels, and goose barnacles (8E). The Pacific goose barnacle, *Pollicipes polymerus* is restricted to the upper two-thirds of the intertidal and unlike its relatives (the smaller *Balanus* and *Chthamalus*) has an extended stalk that enables it to orient itself in the water in the direction of the current flow.

Many species of the low intertidal, such as the brittle star, are not comfortable on exposed rock faces, and must seek out the shelter of crevices, overhanging rocks, or masses of seaweed.

The characteristic seaweed of the lower intertidal is the kelp, *Laminaria* (12E), which provides shelter for a variety of animals including the various nudibranchs or sea slugs (20G, 20H), the large black chiton, *Katharina tunica* (20G), and the black abalone, *Haliotis cracherodii*. The seaweeds attach themselves to rocks with a complex holdfast and this structure provides protection and shelter for a myriad small worms and crustaceans, such as crabs, amphipods, and isopods (6E). The thickened stalks of the *Laminaria* serve as attachments for encrusting organisms such as bryozoans and hydroids (20H), and as homes for more mobile creatures like the keyhole limpets. Many hydroids have a bizarre reproductive cycle, with a fixed, plantlike generation producing free-swimming offspring, whose own progeny are once again fixed.

A characteristic creature of sunlit pools in this zone is the giant green anemone, *Anthopleura xanthogrammica* (9E), which owes its color to the presence of symbiotic algae in its tissues. A species often associated with this anemone is the predatory snail, *Opalia crenimarginata*, which feeds on anemone tissues, and the sea spider, *Pygnogonum stearnsi*.

SEAWEEDS

The plant life of a rocky shore displays a similar zonation to the animal life. Small green algae grow quite high up in the intertidal, but are gradually replaced by red and then by brown algae passing down the shore. The rock weeds *Pelvetia*, common in the high intertidal are replaced by the more turflike *Cladophora* and the flaplike *Iridaea* in the midlittoral zone.

In general, the size of the individual plants increases down the shore, with the largest seaweeds, the kelps (12E) and their relatives being found in the subtidal zone. Air filled bladders, which provide flotation, are found in many of these species, as well as in the common *Pelvetias* and *Fucus* of the intertidal zone. The clumps of seaweeds at different levels on the shore not only provide a source of food for grazing creatures, but their fronds serve as nursery areas for many species, including the octopus (16H).

During exposure at low tide, although the upper surfaces of the seaweeds may become dry and hot, the air beneath them is moist and cool and helps prevent desiccation in the soft-bodied animals that shelter there.

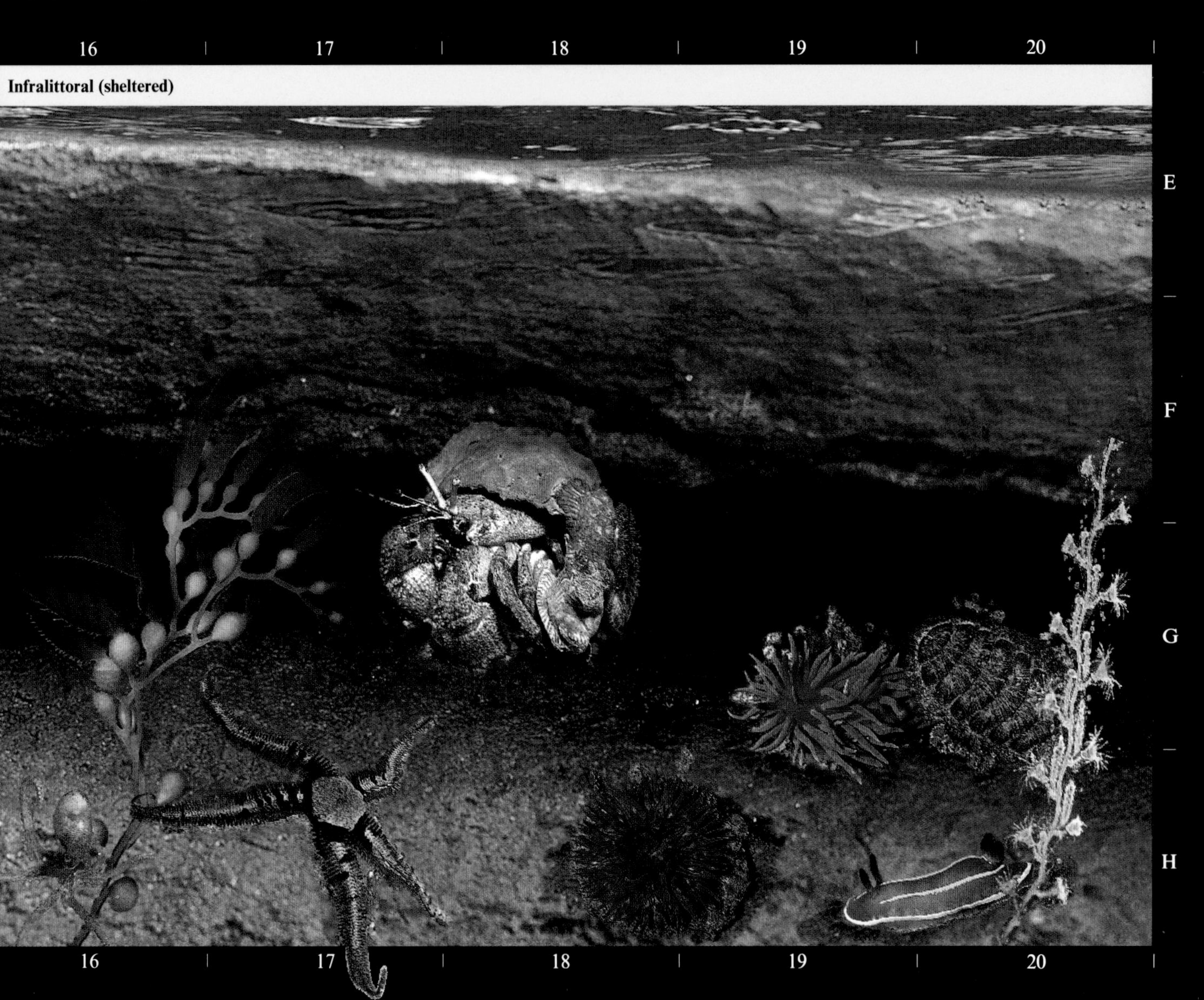

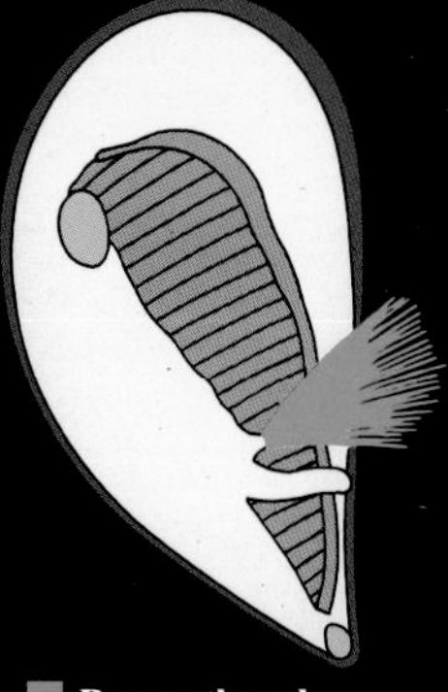
■ Byssus threads

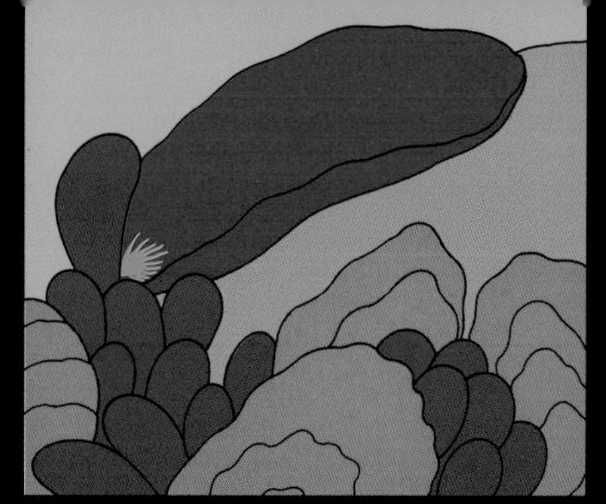
Mussels anchored among oysters

MUSSEL BEDS

Although the distribution of animals across a shore is dominated by the influence of the tides, the degree of exposure and the nature of the substrate, these conditions themselves can be modified by the presence of some animals growing in extreme abundance. The mussels, *Mytilus californianus* (9E), growing on a rock surface alter the wave energy and provide shelter for numerous smaller animals that would not be able to survive there if the mussels had not colonized the surface first. The California mussel bed forms a distinct community that includes the small flattened crab *Petrolisthes*, the isopod *Cirolana* (6E), and nereid worms (8E). The mussels are attached to the rock surface by strong byssus threads that also serve as shelters and traps for detrital material which is fed on by the smaller soft-bodied animals. Byssus threads are incredibly tough strings of protein secreted by the mussel's small, finger-like foot. Immature mussels can even use their byssus threads as guy ropes, in order to help them climb walls.

A well-developed community may take over five years to become established and the mussels themselves may be preyed on by *Pisaster ochraceus* (9E), the common sea star of this section of the California coast. Other mussel predators include the drills or *Thais*, snails that puncture a neat hole in the mussel shell to extract the flesh. The mussels in turn feed by filtering plankton from the current of water drawn into the body by the action of the cilia. During summer months the abundance of *Gonyaulax*, a toxic plankton, in coastal waters results in the mussels themselves becoming toxic.

11 | 12 | 13 | 14 | 15

Infralittoral (exposed)

11 | 12 | 13 | 14 | 15

TENTACLES

The sea anemones and corals are a group of sessile (fixed) animals that feed on plankton and suspended particles caught by the crown of tentacles surrounding the mouth. These creatures are extremely simple: each consists of a single columnar polyp, the central space of which serves as a digestive cavity. Some of the larger anemones may feed on small fish or crabs that are caught and held by special stinging cells that cover the surface of the tentacles.

In the intertidal zone are the brightly colored solitary corals *Balanophyllia elegans* (9D) and *Astrangia lajollaensis* alongside the small anemone *Epiactis polifera*, which is unusual because it does not release its eggs into the water like other anemones, but retains them in brood pits located at the base of the body wall. The young anemones develop in these pits and form an expanding colony, although each individual is independent of any of the others.

Many rock surfaces in the middle intertidal are covered by *Anthopleura elegantissima*, a small delicate anemone that reproduces by budding to form self-contained colonies, and covers itself with sand and gravel when exposed at low tide.

COASTLINE DYNAMICS

The key to success

- Predictable tidal patterns that expose the inhabitants to a regular regime of exposure and inundation
- Animals' and plants' ability to escape the worst of the wave action by fixing themselves to a surface, wedging themselves into crevices, or burrowing into sand or mud

Forces for change

- Rockfalls and general erosion caused by wave action
- Deposition of sand and sediment
- Low river conditions affecting the salinity in estuarine regions
- Outbreaks of predators denuding areas of coast
- Pollution flows from inland sewage works and from offshore spillage
- Overharvesting of the ecologically important mussel beds

Few habitats on Earth change quite so rapidly as coastlines, shifting by the moment as the tide ebbs and flows and one wave after another strikes upon the shore. A coast can change dramatically overnight as a violent storm throws shingle high up the beach, or a section of cliff, long battered by waves, wind and weather, finally succumbs and crumbles into the sea. Or it can change subtly and gradually over the years, as the waves eat away at the coast here and deposit sand there. Such short-term changes are superimposed on much longer cycles of change as the sea level goes up and down with the coming and departure of the ice ages, and the slow, seismic shifts in the Earth's crust. Organisms of the shoreline have adapted to this constantly changing habitat, but each type of change allows a new range of species to flourish.

Shifting shores

Shorelines are shaped by three main processes, each going on in different parts of the coast all the time, or on the same part of the coast at different times. First there is erosion – the continual wearing away of the coast, and the slow breakdown of rocks into pebbles and shingle, sand, silt, and fine mud. There is transportation – the movement of rock debris along the shore or out to sea by waves and the currents. And there is deposition – the dumping of pebbles, sand, and silt to form new sea bed and shoreline features, such as beaches.

It is waves that do most of the erosive work, though they are helped by wind and weather – and by the corrosive effect of salt spray. Storm waves pick up enormous loads of rocks, sand, and pebbles and hurl them against cliffs as they surge towards the shore. It is this constant battery of stones that does most of the damage, but the waves can also ram air into crevices in the cliff-face so hard that the pressure can literally burst the rock apart. As the base of the cliff is undercut, the rock above becomes unstable, and eventually crashes down in a jumbled heap of angular pieces.

The persecuted seal
Seals have suffered badly at human hands, hunted for their fur, culled for their alleged affect on fish stocks, poisoned by pollution.

In this way, the erosive action of the sea opens up a whole new habitat. Plants and creatures unable to gain a foothold on the sheer cliff face find a wealth of suitable homes on the tumbled rocks. Algae and kelp cling to the upper surfaces, along with barnacles, limpets, mussels, whelks, and winkles, while sea anemones, crabs, and sea urchins find shelter from the wind and sun in damp crevices and on the underside of rocks.

Eventually, the sea may wear the cliff back so far that it carves out a rock platform in front of the cliff. The pools of water left behind in this rock platform as the tide ebbs provide permanently wet habitats for a whole new range of plants and creatures. Certain species of the little fish the blenny, for example, spend half their lives in such tidal pools.

Headlands usually bear the brunt of erosion – waves reach them first and so their destructive power is concentrated. By the time they reach bays, however, waves have lost much of their energy and so wash up on to beach all the debris they have worn away from other places on the coast. The steep, shingly, constantly shifting beaches thrown up by winter storms and on exposed coasts provide an unwelcoming habitat, and contain few animals. Very sheltered, muddy shores are equally unwelcoming. Creatures that burrow in beaches, such as clams and polychetes, thrive best on the gently sloping sandy beaches that are lapped by moderate waves in summer.

Longshore drift

Waves seldom roll in exactly parallel to the shore, so the water tends to run up the beach at an angle, then runs straight back down the beach. The overall pattern is a zigzag motion, called longshore drift, which moves sand grains along the beach. In holiday resorts, barriers called groynes are built out from the shore to hold the migrating sands.

Sediment picked up at one point may be dropped again several miles down the coast to form a sand spit or a pebble bank, like the 16-mile long Chesil Beach on the Dorset coast of England. Often a lagoon forms behind the spit, opening up a new habitat for all kinds of wetland plants and animals.

Human changes

To all these natural changes to the coastline are now added an increasing number of human impacts. Coasts have been subjected to heavy industrial development, as nuclear power stations and oil refineries are constructed on the shore for easy access to water. More recently, coasts have been "developed" all over the world for holidays, with hotels, promenades and all the attendant paraphernalia of a seaside resort encasing the shore in concrete.

All this development has had a serious effect on shorelife, especially on sandy shores. Creatures that nest on beaches – sea turtles and horseshoe crabs, and birds such as the piping plover and least tern – have been driven close to extinction by the damage to their nests. Other birds are subjected to the attacks of rising numbers of gulls, which thrive in the rubbish-strewn urban holiday world.

Engineering works to protect the coast deprive many organisms of their natural habitat. Moreover protecting the coast in one place can harm it in another. Spits such as Spurn Head in Yorkshire, England, are maintained with a constant supply of new sand and silt eroded farther up the coast. If the coast is protected, the source of sand may be cut off, the spit is quickly washed away, and an important coastal habitat is lost.

Many coasts, especially those near the world's big industrial centers, are also being assaulted by rising levels of pollution, especially by oil and sewage. The affect of oil spills on birds is all too obvious, but the whole coastal ecosystem suffers. Periwinkles, limpets, and whelks, for example, cannot gain a foothold on oily rocks, and so algae take over, only to be torn away when they too become oil encrusted. The devastation to crustaceans and algae works its way quickly through the whole food web.

Sea otters and kelp

Many coastal creatures have been subjected to direct human attack. Various fish, such as the common sturgeon, have been almost wiped out by fishing, while marine mammals such as seals and sea otters have suffered badly from fur hunters.

The remarkable sea otter was hunted almost to extinction not only for its pelt, but also by fishermen who saw it as a threat to fish stocks. Yet its elimination could threaten the survival of one of the richest of all coastal habitats, the undersea kelp "forest."

Huge beds of giant kelp seaweed grow off many rocky shores, especially off the west coast of North America, and provide a unique habitat which is used as a breeding ground and nursery by many fish species. This seaweed also provides a home to the sea otter, which plays a special role in its maintenance, and to the giant sea otter, which helps to keep down the numbers of sea urchins, whose voracious feeding could otherwise cause devastation to the kelp.

The agile otter collects a stone from the sea bed, floats at the surface with the stone clasped to its chest, and uses it as an anvil on which to break open the tough shells of sea urchins and abalones. It is now realized that the otter is one of the few effective predators on the sea urchins that, in some areas, have caused the destruction of the kelp beds. Kelp is now a protected species, yet numbers have declined so much that it is still extremely vulnerable.

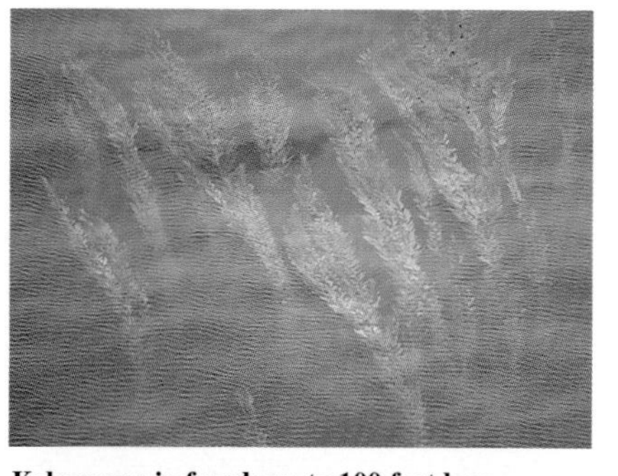

Kelp grows in fronds up to 100 feet long

A resting otter wraps itself in kelp as an anchor

Unchecked by the sea otter, sea urchins can destroy a kelp forest entirely

DRY MIGRATIONS

Many of the larger animals and birds migrate in response to seasonal changes. Elephants (10B) from the Okavango travel fairly short distances, in comparison with some species, but their movements follow a definite pattern. During the dry season, when the surrounding country is barren and parched, the herds gather at the swamps which have recently been swollen by floodwater from the distant mountains. They are not the only grazers to compete for the lush marginal vegetation, however, for many others use the same dry-season refuge. When the rains arrive at the end of the year, the elephants return to the woodland, which now has numerous "pans" filled with water.

Outside the permanent swamps, in which conditions underfoot are difficult, lions (10A) are major predators. During the rains, many of them shadow the herds into the Kalahari, but as the following dry season begins they trail their prey back to the Okavango, there to fight with the prides that have remained behind for a stake of the prime territory.

Many waterbirds undertake migrations that may be both extended and complex. Abdim's storks breed north of the Sahara and winter in southern Africa. Many other species breed in the Delta and time their arrival so that most of the nest-building and egg-laying is completed before the rains. The chicks then hatch out when water levels are lowest – and when the density of fish is therefore at its highest.

But only a minority of species seasonally travel away from the swamps. Animals such as the reed frog (2D) and the cat-eyed snake (5D) are poorly adapted to life outside wet areas. Many insects, notably dragonflies (3C), have wings capable of carrying them great distances, but this is a random process and does not involve any regular migratory pattern.

LOW OXYGEN

Oxygen levels in water tend to be very low where the swamp vegetation is thickest, because the shading reduces photosynthesis, there is little mixing of the water, and the decay of plant material itself uses up oxygen. Some catfish swim to the surface to take in air, and certain semiaquatic snails have built-in "snorkels."

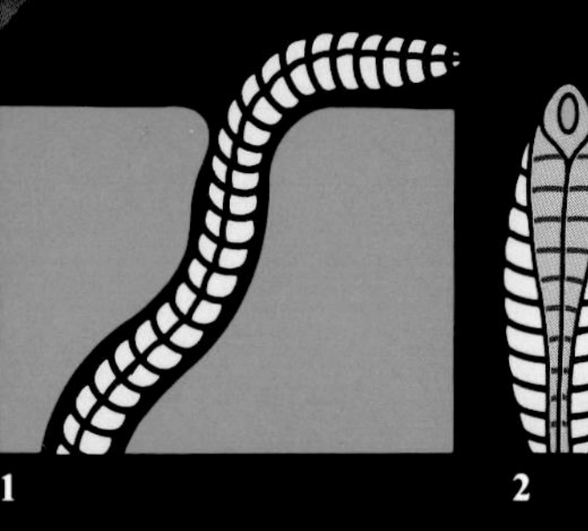

1 2

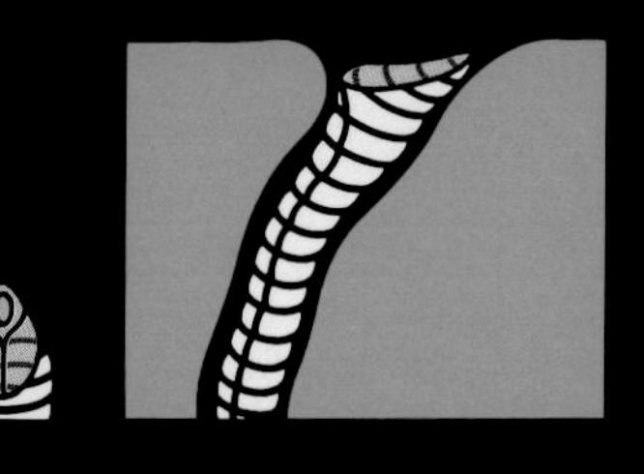

3

The swamp worm *Alma emini* (1) is abundant in low-oxygen mud. To obtain oxygen, it exposes its hind end, which is rich in blood vessels and folded to form a temporary tubelike "lung" (2 and 3) through which bubbles of air are drawn. Most of the oxygen is withdrawn from these bubbles and is passed to the tissues by the highly efficient hemoglobin in the animal's circulation.

For a "dry" continent, Africa supports an amazingly large area of wetland – some 4 percent of the total land area. Many African swamps are additionally subject to periodic flooding. The Okavango Delta in Botswana, for example, has two flood peaks a year, one as a result of local rainfall, and the other a result of the wet season in the headwaters of the Okavango River in Angola. So an area of wetland which at low water comprises some 4,250 sq miles of permanent swamp can swell at times to cover 11,000 sq miles.

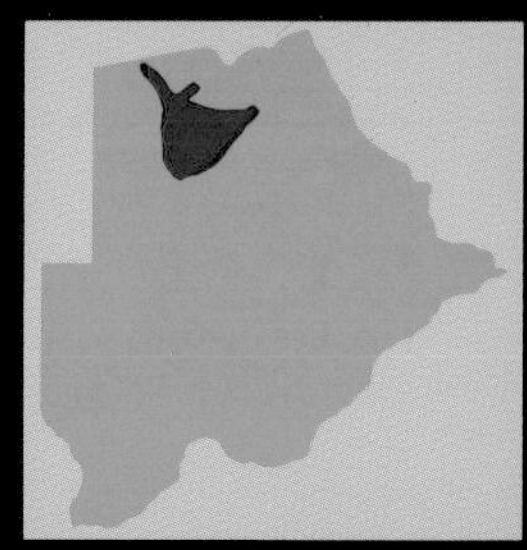

THE CATFISH RUN

Migration is not only a prerogative of the land animals. Seasonal travel also takes place under water. Between 70 and 80 species of fish are recorded in the Delta, and the movement and spawning of most of them is stimulated by the floods. Inundation periodically connects the water bodies on the floodplain to the main channels, making available to the fish a huge amount of food on what was land. It also opens up excellent spawning sites, and tilapia, in particular, undertake extensive migrations in response to this phenomenon.

When the level of the Okavango River is at its lowest, a boiling and splashing commotion at the surface is frequently to be observed, drawing the attention of fish-eating birds, crocodiles and snakes. It is a sign that the annual "catfish run" is under way, an event that involves the upstream migration of shoals of sharp-toothed and blunt-toothed catfish (6D), each of which may be well over a yard in length. Like an army on the move, they attack and consume smaller fish on the way, thrashing against the papyrus plants to drive their quarry out into the open. This event is unique to the Okavango, and allows the catfish to carry out intensive feeding at low water, when prey is at high density. They then move out onto the northern floodplains, an area suitable for spawning, during the floods in January and February.

Tiger fish – also fierce predators – undertake a similar migration, benefiting in part from the availability of small fish that have been flushed out by the pack-hunting catfish.

SWAMPS, BOGS, AND MANGROVES

Wetlands occur all over the planet – from the polar regions to the tropics, along the coasts and in the very center of the continental landmasses. They may occupy vast areas, or form islands within dry landscapes. Diverse in nature, all wetlands, nonetheless, have features in common – the more obvious of which is that their biological, physical and chemical traits have both aquatic and terrestrial components.

Their diversity, however, makes classifying wetland types complex, and terms used have a range of meanings: the word swamp may be used as a rough synonym for wetland, or as here, to describe an area that is permanently or often flooded. Bog here means an area with waterlogged soil, where there is usually little standing water, and where peat tends to accumulate. Mangrove swamps have mangrove plants, mainly in tropical latitudes, in ecotones (areas between differing adjacent floras) influenced by tides.

Wetlands form a sort of bridge between land and water, although the balance between the aquatic and terrestrial elements is not fixed, and the fauna and flora of wetlands must be able to survive continual changes, from complete immersion in water to desiccation. Evolutionary methods by which organisms cope with such fluctuating conditions range from the cocoons in which African lungfish lie buried in order to survive the dry season to the long-distance seasonal migrations of birds.

Some plants, like sedges, are found in wetlands all over the world. Others like mangroves in the tropics are found only in certain regions. Plants that grow on sites which are permanently or seasonally flooded, or in which the soil is normally waterlogged, are known as "hydrophytes." Each has its own way of ensuring a steady supply of air to the roots. Mangroves grow modified roots, or pneumatophores, visible above the mud at low tide, which absorb oxygen from the air and transport it to the buried parts. Other hydrophytic plants, such as water-lilies, have stem and leaf tissues that consist of thin-walled cells separated by large air-filled spaces. This conveys air to the roots but also provides the stems with buoyancy in water.

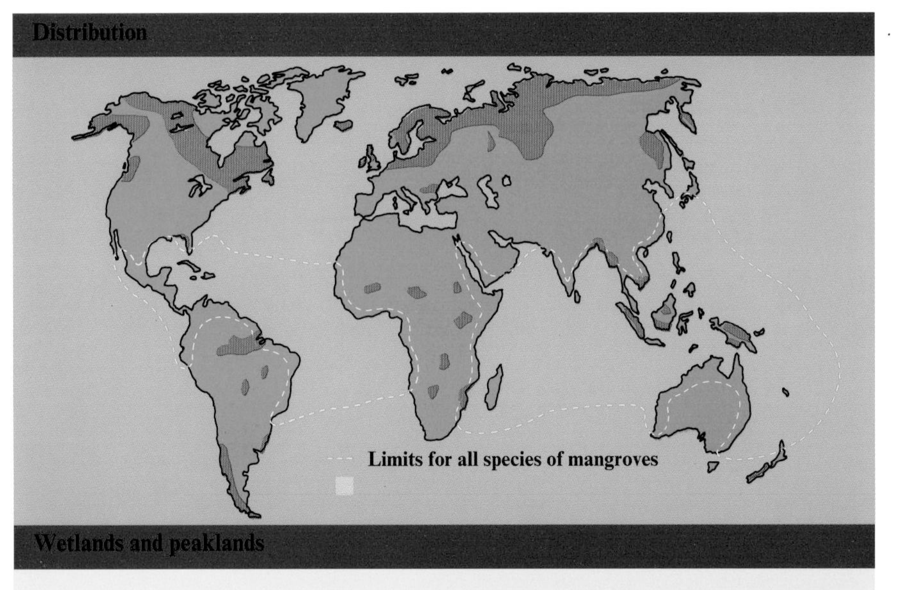

Wetlands and peaklands

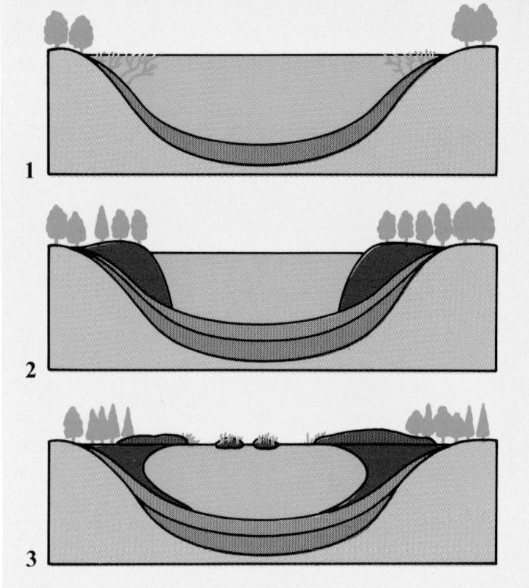

Bog formation
Peat bogs form in cool, damp regions as lakes are slowly filled in over a period of 5,000 years or so. As lake sediment collects around aquatic vegetation (1) the lake margins turn to swamp allowing trees such as willow and alder to colonize (2), but the trees are soon replaced by sphagnum moss and, in steep-sided lakes, by floating carpets of sedge, which slowly turn to peat (3).

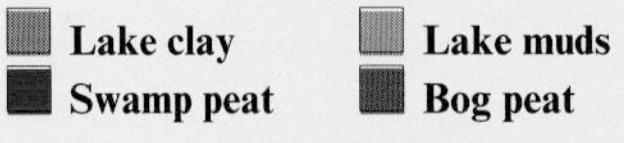

Changing vegetation

Differing vegetational zones can usually be distinguished along a line known as a hydrosere, running from dry land to the heart of a wetland. On the margins, for example, the more common plant species tend to be those adapted to drier conditions, whereas in the wetland center truly aquatic plants ordinarily prevail.

Once plants are established in wetlands, sediment often begins to build up, and the ecosystem goes through a succession of plants, adapted to the increasingly dry conditions. Typically this begins with floating plants like the waterlily, followed by emergent species, which root in the shallows.

The emergents grow thick and fast, and biomass productivity can match that of tropical rain forests. In many African lakes, the emergent is papyrus, which often breaks away to form floating mats of vegetation. In the temperate zone, there is often just a single emergent such as the reed mace or saw grass. Trees, such as the swamp cypress of the Florida Everglades, may grow with the base of their trunks always in water.

The creatures of the wetlands

Wetland creatures may be either land or water living, or equally at home in both. On the floodplain of a swamp, for example, great schools of fish may move in during periods of high water, but as the water recedes their place may be taken by herds of grazing mammals. The larvae of many insect groups, such as the dragonflies, develop in the water and then emerge into the world above as adults.

Wetland mammals are generally small, with the notable exception of the

The trees of the Everglades, which cover some 6,500 sq miles of southern Florida in the United States, grow with their roots perpetually submerged

hippo in Africa. In South American swamps, the largest mammals are rodents, the capybara and coypu.

Seasonal changes have dramatic effects, and may also be extremely important in maintaining the stability of the ecosystems. In swamps, for example, the migratory cycles of many fish, birds and mammals are regulated by seasonal floods, which may also bring rich deposits of silt. In areas of tundra in the Arctic, bogs are frozen over and snow-covered for much of the year, but life bursts forth during the brief summer months. The pattern of high and low tides, related to the lunar cycle, is most important in mangrove forests, for different species occupy zones according to the frequency of immersion by the tide.

But it is for their birdlife, above all, that wetlands are renowned. Short-legged wading birds like the avocet probe the mud close to the shoreline; long-legged wading birds like the heron hunt farther out; farther out still, birds catch fish by surface snatching or by diving; and in the air above, birds seize insects by the beakful. Mangrove swamps have the spectacular scarlet ibis and fish-eaters such as the anhinga, while temperate marshes have the marsh harrier, and the Everglades the Everglade kite, both which prey on small mammals and other birds.

Humans and the wetlands

In recent years, wetland areas have been under increasing threat from human activities. They were protected in the past by such factors as inaccessibility, a human social disposition that incorporated a respect for (even a wonder at) the natural world, smaller human populations and the local presence of diseases like malaria. People such as the Marsh Arabs of Iraq and the Dinka of Sudan enjoy cultures that encourage them to make sustainable use of wetlands. But these two groups have recently fallen victim to violent supression by governmental authorities. Today, big business – such as the prawn farms responsible for the removal of much of Southeastern Asia's mangrove swamps, such as the agricultural enterprises worldwide that drain swamps to produce new farmland – have destroyed much of the planet's most precious wetland. In many cases these projects bring only short-term gain for the minority while eliminating the livelihoods of local inhabitants or nomadic communities.

Ecological counterparts

Species of crocodiles and alligators fill similar ecological niches in swamp and mangrove ecosystems. The (North) American alligator, the caimans from Central and South America and the Nile crocodile from Africa are all similar in appearance; the gharial from India has a snout which is specially adapted for fishing.

Crocodiles and alligators existed with the dinosaurs, having undergone little change since. Their streamlined shape and their strong, flat-sided tails allow rapid swimming. Eyes and nostrils are set high on the head so that they can see and breathe when almost totally submerged, allowing for stealth, while camouflaging the young from predators. Common quarries are amphibians, fish and mammals.

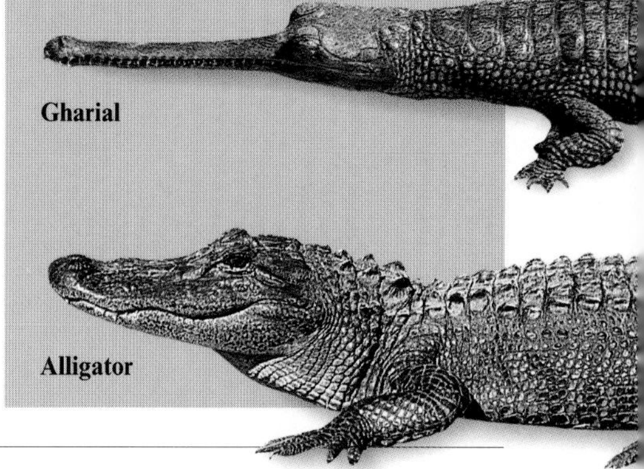

BIRD LIFE

More than 400 species of birds have been recorded in the Okavango Delta, but bird life is relatively sparse in the area of permanent swamp, partly because of the lack of sites suitable for waders. As is the case in relation to many animal groups, the margins of the permanent swamps are where birds are most abundant and diverse.

The Delta is a refuge essential to many species of endangered birds. The northern part, for example, is one of only three breeding sites for pink-backed pelicans (8D) in southern Africa.

A remarkable diversity of bills is apparent in birds that snatch their prey from the water, ranging from the scoop-shaped arrangement of the pelicans to the flattened form of the African spoonbill (17C). Almost every conceivable niche is exploited. Insects are taken in the air by bee-eaters, kingfishers dive for small fish, and oxpeckers pick food from the backs of large vertebrates such as hippos.

SURE FOOTING

Many animals adapted for life on solid ground are ungainly when confronted with a surface that sinks beneath the feet, and that may be covered in water. Animals that are adapted to the wetland environment, however, can move at speed on such a surface, and thus have a great advantage over land-based predators such as lions.

The antelope known as red lechwes (11B) have elongated hooves which, together with their bounding gait, make it possible for them to run faster through shallow water than on dry land. The hooves of the sitatunga (6B) splay widely to support the animal's weight on boggy ground.

Even comparatively weighty birds avoid sinking into the mud thanks to elongated and widely separated toes that spread the load over a greater area.

Wetland reptiles have no weight distribution problems: a large proportion of their bodies is in contact with the substrate at all times.

16 17 18 19 20

A

B

C

D

16 18 20

swamps with rooted vegetation, for example, contact with the bottom means that the sediment may represent a source of nutrients, and the network of channels or lagoons that is usually present allows better mixing of the water.

In the Okavango the succession of plant types across the floodplain, from deep-water species to those only adapted for occasional flooding, is far from smooth because the area is scattered with islands, each of which has vegetational zones. Even the termite mounds – an important feature of the area – provide sites for colonization by woody species that could not survive long-term flooding.

- Autotrophic layer
- Heterotrophic layer
- Floating mat
- Peat
- Sludge and detritus
- Sediment and mud

AIRBORNE PREDATORS

The largest of the Delta's aerial hunters are the African fish eagle (15A) and the mainly nocturnal Pel's fishing owl, about which little is known – not even how it locates its prey at night.

In spite of the specialized niche that the fish eagle occupies, it is similar in general appearance to its land-based relatives. It does, however, have sharp spines on the undersides of the toes, which allow slippery fish to be gripped firmly. The fishing owl, similarly, has sharp-edged spiny scales on its unfeathered feet. Perhaps because fish are insulated from sounds above the surface, many fish-eating raptors lack plumage adapted for quiet flight.

The fish eagle usually snatches its prey from the surface, but is also known to dive right into the water, driving itself back up and out again with powerful wing-strokes. Heavier fish, up to 3.5 lb, are ordinarily dragged along the surface of the water before being heaved up at the bank. Hunting most often begins from a perch, but reconnaissance runs are also undertaken. When a fish is sighted, the eagle descends slowly, before making a feet-first plunge.

The courtship of African fish eagles involves aerobatic displays in which a pair tumbles toward the ground with talons locked together, only releasing and soaring away just before impact.

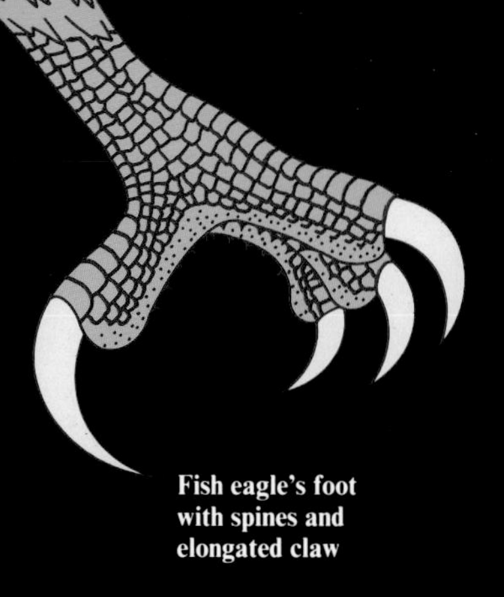

Fish eagle's foot with spines and elongated claw

11 | 12 | 13 | 14

11 | 12 | 13 | 14 | 15

HIPPOS AND CROCODILES

The two largest permanent inhabitants of the Okavango swamps are fierce fighters, although one – the hippopotamus (14C) – is vegetarian, and the other – the Nile crocodile (1C) – gently carries its young between daggerlike teeth. Both occupy the same type of habitat, but there is little competition. Indeed, the trampling around of the hippos in the papyrus and reed beds helps to create suitable nesting sites for the crocodiles. Hippos additionally play an important role in the cycling of nutrients: they graze on the exposed floodplain at night and return to the water by day, where they continue to defecate. In this way, a group of 15 hippos may transfer about a ton of biomass to the water each week.

Okavango crocodiles are significant predators of fish and, when older, mammals. Their breeding sites are located mostly on sandbanks in the northern part of the Delta, from where many of the young travel downstream.

SWAMP VEGETATION

Swamps in Africa may be reed or forest, rooted or floating, seasonal or permanent. Several types may occur in one body of wetland. In the Okavango Delta, the core is a permanent swamp where water is up to 13 feet deep.

Much of the deeper water is covered by floating beds of papyrus (6B). Rooted plants such as hippo grass (15C), reeds, waterlilies (18D) and bulrushes flourish in shallower parts. In places, raftlike segments constantly break away from the main papyrus beds and are carried downstream.

In both swamps with rooted and floating plants, the bulk of primary production takes place by means of photosynthesis in the autotrophic layer, which is near and above the water surface. Below this is the heterotrophic layer, where most of the consumption and decomposition occurs. Despite this similarity, the two swamp vegetation types exhibit a number of different environmental characteristics. In

SWAMP AND MANGROVE DYNAMICS

The key to success

- A balance between the terrestrial and aquatic elements of the system
- A regularly replenished water supply, either from local precipitation or brought from a distance by rivers or aquifers

Forces for change

- Major changes to the drainage pattern of the area
- Fire
- Felling of trees in and around the wetland for timber or charcoal
- Chemical pollution
- Introduction of new animal or plant species
- Meteorological extremes – flood or drought
- Change in the salinity of the water
- Animal incursions from neighbouring, drier regions to escape a period of drought

All wetland ecosystems are adapted to continual changes in their environment. So many changes, rather than disrupting the ecosystem, only serve to increase its stability and boost productivity. In a mangrove swamp, for example, changes occur twice-daily as the tide ebbs and flows, fortnightly with the spring and neap tides, and seasonally according to the weather.

Once plants have colonized a wetland, their presence can stabilize the ground by increasing the amount of material trapped or deposited. The waterlogged conditions inherent in a peat bog, for example, result in the very slow decay of organic matter and so the remains of dead plants accumulate to form peat. These are perfect conditions for colonization by bog moss (*Sphagnum* spp.).

Nevertheless, the plants and animals that inhabit wetlands do not form the close web of mutual interdependence that is typical of tropical rain forest communities, for example. This may be because many of the animals are visitors, such as migratory birds and fish. In mangrove swamps, dispersal takes place almost entirely by water. So there is little or no evolutionary advantage in producing fruits and seeds attractive to animals as rain forest trees do.

Drainage changes

On a large scale, it is almost certainly shifts in drainage patterns that cause the greatest disruption to wetland ecosystems. Such shifts may turn areas of remarkable beauty and diversity into monocrop deserts or anoxic lakes.

The drainage may be disrupted by natural forces such as earthquakes or silting up, but human activities exert a huge influence too.

Sometimes, this is part of a deliberate attempt to reclaim new land for agriculture or aquaculture. Farmers used wind-powered drainage mills to dry up the fens and broadland rivers of East Anglia, for example, as early as the 16th century, and even earlier in The Netherlands. But modern machinery and the increasing demand for food has led to a recent acceleration in the draining of wetland areas.

Drainage patterns may also be altered by humans for other reasons. The Jonglei Canal scheme in Sudan, for example, sought to divert half the flow of the White Nile away from the Sudd swamps in order to provide more water for downstream consumers. The scheme was resisted by local people who stood to lose valuable resources; worldwide press coverage gave publicity to the threat to wildlife. An ongoing civil war has since caused suspension of the project.

Fire damage

Small-scale changes to wetlands may not necessarily be for the worse, however. In the bogs of Scotland and Ireland, peat has for centuries been cut for fuel. Where small pools are created in abandoned workings, the habitat diversity is increased.

The mudskipper

Few creatures are better adapted to the boundary between land and water than mudskippers. They are fish that live in swamps and mudflats all around the Indian and Pacific Ocean, from Africa to Polynesia. Yet they can quite happily climb, walk and skip about out of water whenever the sea recedes. When they are out of water, they breathe air that is trapped within their gill chambers.

Another cause of disruption to wetlands is fire. Deliberate burning by humans is probably the most common way in which fire breaks out in wetlands, although lightning also has a significant effect. On many of the seasonal floodplains in Africa, for example, the vegetation is burned at times of low water in order to encourage the growth of new shoots on which cattle can graze.

Where the wetland vegetation is in the form of stands of trees, there remains the threat that the trees will be felled for timber or to make charcoal. Sometimes – but not always – felling may be the precursor to the conversion of the land for farming.

Pollution poses an increasing threat to wetland communities. Many of the areas of mangrove and saltmarsh are in river estuaries, locations that tend to attract industrial development, the waste products of which may then be discharged into the estuary. Polluted inflows and acid rain similarly affect swamps and bogs.

Animal introductions

Introduced plants and animals can also cause ecological disruption, for they may spread rapidly and bring about major changes to the habitat. Recently a large herbivorous water rodent from South America called the coypu has been introduced into North America, Europe, and parts of Asia, for instance. Wherever it arrives, the coypu has thrived and the grazing pressure on native wild plants and on farm crops has seriously increased.

The long term stability of the wetland ecosystem is deeply susceptible to the influence of the weather. Hurricanes, and their attendant incredibly high winds and tidal waves, can devastate mangrove ecosystems – and just a small change in the world's temperature patterns could alter their distribution and frequency. Shifts in rainfall patterns, of the sort that cause deserts to expand, may likewise reduce the area of freshwater wetland.

Mangrove swamps

Natural variations in salinity occur in mangrove swamps where sea water may be further concentrated by a rise in evaporation or diluted by river- or rainwater as rainfall rises. High salt concentrations make water uptake by the vegetation difficult, and mangroves have had to evolve unique survival mechanisms. Like many desert plants, they

Mangrove zonation

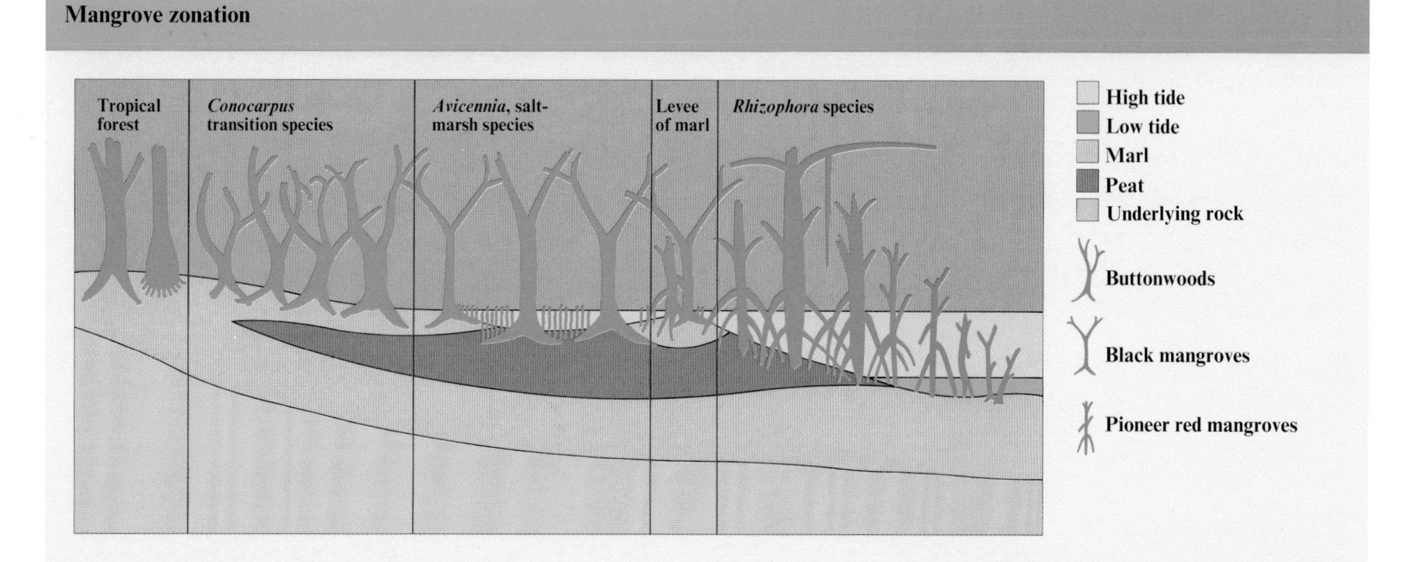

Mangroves: roots, fruits, and zonation
Like many wetland communities, mangrove swamps often show a distinct zonation of vegetation from the low-water level landward (**below left**), reflecting the different plant species' varying tolerance of inundation and desiccation.

In a typical Florida swamp, for example, red mangrove (*Rhizophora mangle*) (**below right**) – a pioneer species – is the mangrove best able to survive frequent and deep flooding, and so this plant dominates the lowest zone. Closer inland, the vegetation is characterized first by black mangrove (*Avicennia germinans*) and then by button mangrove (*Conocarpus erecta*). At the rear of the mangrove swamp there may be areas of freshwater wetland or raised ground with tropical forest.

The zonal pattern may not be obvious immediately, especially when hummocks of upland – on which terrestrial forest trees grow – are scattered throughout the swamps. In forests that are often hit by hurricanes, red mangrove comes to predominate because it survives better than black mangrove.

Fewer than five mangrove species occur in the Americas, but a far greater diversity occurs in Southeastern Asia and parts of Africa, where the mangrove wetlands may contain more than 20 species but these tend to display similar zonation.

have leaves that are succulent and able to excrete excess salt, which then drips from the leaves, especially when washed by rain. Because mangrove species differ in salt-tolerance, variations in salinity in swamps result in differences in community structure.

It is the mangrove's remarkable roots, however, which enable it to colonize areas of unstable and oxygen-deficient mud. Red mangrove species, for example, have a mass of stilt-like arching roots that provide both firm anchorage and a means of exchanging gases with the air. Black mangrove species, on the other hand, have roots that stick out of the mud like a bed of nails, but these, too, allow the plant to obtain oxygen from the air. The mangrove's intricate root network encourages silt deposition and so begins the slow seaward advance of the forest, creating new land. The roots also provide attachment points for coastal organisms such as algae, barnacles and mollusks.

Other features of the mangrove which make it such a successful colonizer of tropical coasts are its fruits and seeds. Many mangrove species have seeds which develop on the tree after fertilization. The tree drops a part-grown seedling into the mud. Once the seedling settles or floats off to another mud bed, it immediately sends out side roots to stabilize it.

35 years farmers in the southwestern high plains have tapped more than 375 cubic miles of water. The high plains have as a result become some of the world's most productive farmland – but the aquifer's waters are largely nonrenewable, and are already beginning to recede.

Within the next two decades wells will have to be drilled deeper and deeper, making irrigation prohibitively expensive and reducing the number of irrigated acres from 6.9 to 2.2 million.

BOOMING GROUNDS

In an ecosystem as flat and as exposed as the prairie most animals have had to become masters of subterfuge. Coyotes (9B) blend with the grasses as they slink up on their prey. Horned toads mimic patches of gravel as they bask in the sun. Within so secretive an environment, the flamboyant courtship rituals of the greater and lesser (7C) prairie chicken come as something of a shock. Every year, in the early spring, male prairie chickens gather at what are called booming grounds – named after the basso calls emitted by greater prairie chickens – to cavort in front of female onlookers. Raising feathers and inflating pink sacs at their necks, uttering unearthly noises and performing feats of ungainly gymnastics, these grouselike chickens compete against each other to prove themselves worthy mates.

Unfortunately, prairie chickens – like bison – called too much attention to themselves in the 19th century, and proved too easy to kill. Prized in eastern restaurants for their tender meat, they were all but hunted to extinction. There are today some 50,000 birds, a mere fraction of the millions that existed before European settlement, but their populations are protected and at last their numbers seem to be on the increase.

the armadillo (10C). Although mostly thought of as a South American animal, one of the 20 species of armadillo, a nine-banded variety, has found its way into North America. Despite the armored plates, which cover even its head and tail, the armadillo chooses to live underground for protection, emerging at night to hunt for termites and insects. It will also eat vegetation and carrion. When frightened it rolls into a ball.

DUNG BEETLES

Although often ignored or reviled, insects are cornerstones of the prairie ecosystem: they spread seeds and pollen, break down plants, fertilize the soil, and provide food for birds and small mammals. Not quite an inch long, the dung beetle (or tumblebug, as it is frequently called)(9D) uses its scooplike head to roll a ball of dung sometimes as large as an apple. Once satisfied with its compacted creation, the beetle buries it, feeds on it, and then lays its eggs in it. When the larvae hatch, they finish off what remains of the ball. In this way dung beetles assure themselves of a reliable diet and, inadvertently, distribute seeds that may be rolled up within the dung.

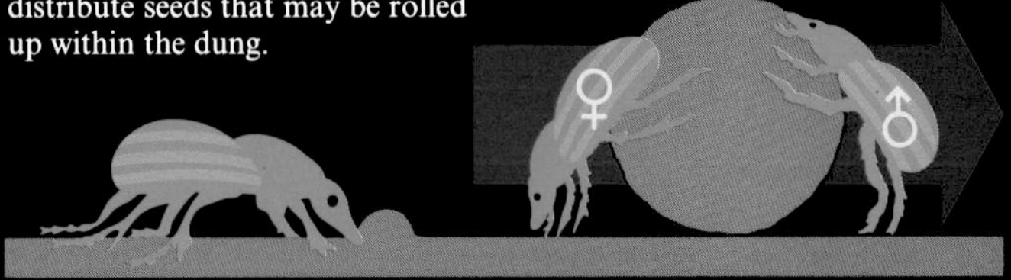

Dung collection and transportation

Growing where once there was a vast Cretaceous sea, the North American prairie undulates across a little more than 247 million acres of land between the forests and the mountains that flank its western and eastern borders. In the west, where the Rocky Mountains' rain shadow ensures an arid climate, the shortgrass (1B) prairie plays host to western diamondback rattlesnakes and other drought-tolerant species. In the east, plentiful rainfall sends forth lush tallgrass (14B), home to hundreds of wildflower species.

THE OGALLALA AQUIFER

The great wellspring of modern American agriculture, the Ogallala aquifer lies beneath 279,400 sq miles of land belonging to eight States in the central plains. Some 10–25,000 years ago glacial meltwater from the Rocky Mountains filled the aquifer with more than 2,300 cubic miles of water. This water trickles deep below the surface through a sandy gravel-bed 150–300 feet thick. Over the last

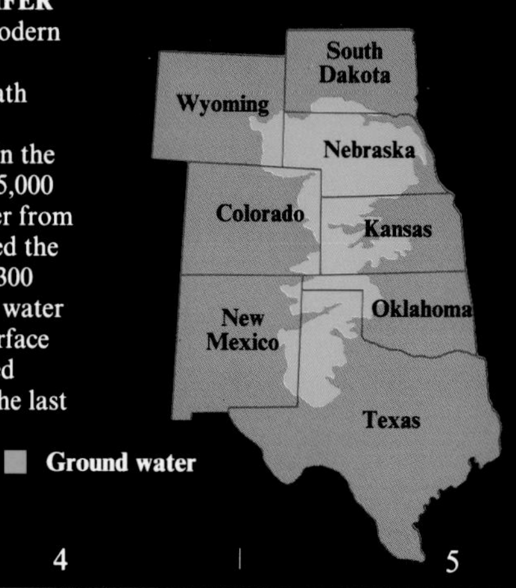

PRAIRIE DOG BURROWS

Prairie dogs (6D) on the high plains alone once outnumbered the entire population of humans on the planet. Their deep burrowing and their grazing helped to aerate the soil and to keep certain grasses in check – areas with prairie dogs typically support more diverse plant life than areas without them. Colonies can stretch for hundreds of square miles beneath the plains. Hawks (17C), coyotes (9B), and other prairie predators gather near them to hunt for small game. Burrowing owls (5D) rear their young in burrows that have been abandoned. Adult owls lay their eggs in the spring and fledge their young in July, at which stage the prairie dogs often reappear and live alongside them. Rattlesnakes (8D) also use the tunnels – but if a prairie dog smells one it tries to seal off the tunnel in which the snake is hiding. Unfortunately farmers and ranchers, convinced that prairie dog tunnels destroy crops and cripple livestock, have tried to eradicate them.

Another burrowing animal is

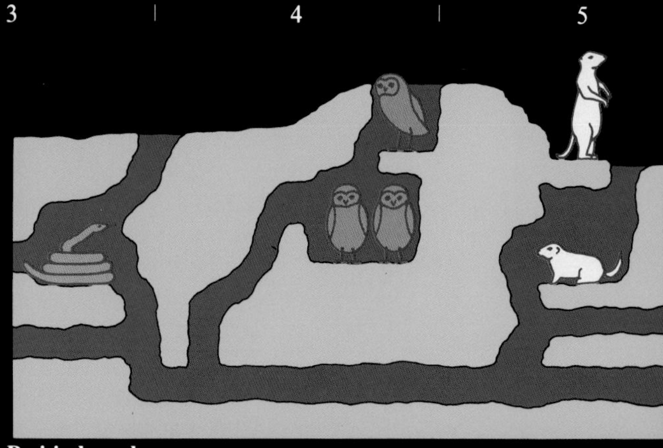
Prairie dog colony

TEMPERATE GRASSLANDS

Dry, flat, and seemingly unending, the temperate grasslands of North America and Asia long appeared as bare and monotonous as oceans. Only late in the 19th century did settlers going west in North America discover that the grasslands, seen close up, were astonishingly fertile and the home of an endless variety of plant and animal life. Beneath waving seed-heads of grasses that could reach higher than the pioneer wagons, the most luxuriant prairies harbored more than 80 mammal species (including bisons by the million), 300 species of birds and many hundreds of species of plants.

The Asian steppe and the North American prairie contain a range of different ecotypes within their margins. As annual rainfall decreases from east to west in North America, and from west to east across Russia, the types of grass species gradually change, along with the animal species they host. Botanists divide the prairie into three broad categories: shortgrass prairie, in which drought-resistant plants do not grow more than half a yard tall; mixed grass prairie, with grasses growing between 20 and 60 inches high; and eastern tallgrass prairie, in which grasses grow up to 10 feet tall, in areas so moist that prairie fires alone keep bordering forests at bay. Shortgrass prairie contains similar grasses throughout its range, whereas tallgrass replaces tropical grasses with temperate grasses as it extends north into Canada.

In Asia, the Altai Mountains divide the steppe into eastern and western ranges. The eastern steppe is higher and drier, and home to the most extreme seasonal temperature variations in the world. Botanically, the steppe can be divided broadly into western forest steppe, in which stands of oak, pine, and birch forest alternate with patches of fescue and feather grasses; open steppe, in central Russia, totally absent of trees except along rivers and their floodplains; and southern semi-desert, where steppe grasses gradually give way to sage brush and true desert vegetation.

Timeless and imperturbable as they may appear, the steppe and the prairie are ecosystems in constant flux, both in the long and the short term. Less than 10,000 years ago, mammoths, camels, sloths, and lions thrived in central North America, grazing and hunting in a park-like landscape interspersed with marshes, savannas, and coniferous forests. As the climate gradually became warmer and less stable, heterogeneous parkland was replaced by hardier, more homogeneous ecosystems: boreal forests to the north, deciduous forests to the east, grasslands and deserts to the west.

Under the influence of bison and pronghorn antelope, North America's grasses changed character. Numbers in the bison herds swelled to the millions – and prairie grasses evolved extremely strong root systems to withstand their sharp hooves and voracious appetites. Periods of severe drought in the west forced the evolution of shortgrass species that could store nutrients in modified stems or deep roots to take advantage of groundwater reserves. Perennial plants evolved shoots and runners in order to propagate in years when few seeds would germinate; annuals such as sunflowers developed massive seed arsenals to improve their reproductive odds.

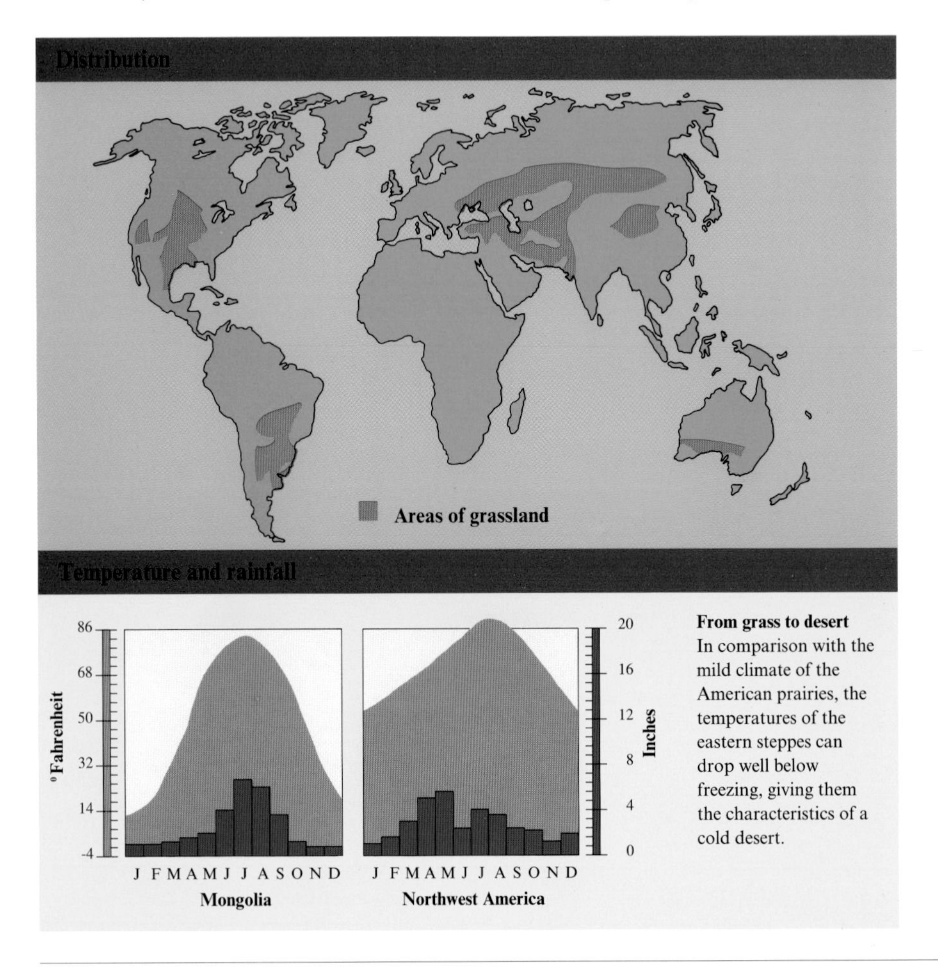

From grass to desert
In comparison with the mild climate of the American prairies, the temperatures of the eastern steppes can drop well below freezing, giving them the characteristics of a cold desert.

The wet-dry cycle

As the prairie and steppe climates today undergo cyclic periods of drought and relative humidity, reproductive strategies succeed and fail, constantly changing the grassland's face. During years of quantitative rainfall on the prairie, the eastern woodland sends seedlings west into the tallgrass which encroaches on the mixed grass, and the shortgrass dispatches runners into the desert. Annual plants, meanwhile, germinate a higher proportion of their seeds, winning ground from perennial grasses.

Studies of tree-rings have shown that a succession of wet years on the prairie is followed, every 22 years or so, by severe drought. This means that the desert and shortgrass prairie take back some of what they have lost to the rain, migrating east across tallgrass and woodland, while perennial grasses gain an advantage over seed-dependent annuals. Fire at this point takes a hand. Known as "red buffalo" to the Plains Indians, prairie fires sparked by lightning can sweep over millions of ácres, clearing the way for fire-resistant perennials, and fertilizing the ground with the ashes of once-dominant annuals.

In spring the prairies become a blanket of bright flowers surrounding the occasional isolated tree

Winds across the prairie

Winds can whip up grassland fires as high as houses, but they also sculpt the landscape by less violent means. North America, in particular, is laid out like a giant wind generator. When dry, cold Canadian air rolls down the eastern slopes of the Rocky Mountains and careens into a languid mass of humid air rising from the south, tornadoes, hail, and some of the world's most powerful storms spin off from the aerial maelstrom. Winter winds regularly reach speeds of 60 mph, creating wind-chill temperatures far below zero. Spring storms can dump over three feet of rainfall in a matter of hours. To withstand such violence, plants on the prairie must bend, hang on, or get out of the way. Most do something of all three by means of flexible stalks, strong roots, and relatively low profiles. Today's crops, poorly adapted to the prairie lands they have come to dominate, do not fare as well. Wheat farmers occasionally waken to find entire fields flattened by errant gusts.

Wild camels on the Mongolian steppe: the ecological counterpart of the American bison

Diversity in flora and fauna

Grasses may dominate the prairie and the steppe, but a sharp eye can still discern the streams, the scarp slopes, the marshes, and the rocky outcrops that divide and embroider the expanse, and that provide myriad niches for animals. Cottonwood, hackberry, and sand plum trees that line the rivers and creeks harbor muskrat, water snakes, and plants more at home in distant forests. Shallow sheetwaters in the tallgrass and mixed grass fill with the first spring rains, becoming terrestrial tidepools where migrating cranes and geese can gorge on frogs, crickets, and water bugs to prepare for their long flight north. Patches of deep, sandy soil with small creeks – home to clusters of trees – give shelter to songbirds and small mammals. Although the sheer scale of the steppe and the prairie can overwhelm human perceptions, a grassland's rich life reveals itself in such details.

Although not directly related, the North American coyote, the Eurasian jackal and the African wild dog play much the same role in their grassland ecosystems. All are largely nocturnal, hunt in packs, and rely both on fresh game and on scavenged meals for their sustenance. Built low to the ground – the better to slink through grass in order to surprise the prey – and capable of sudden, swift movement but not lengthy runs, they are admirably adapted to their lifestyles, and continue to thrive despite human interference. All three are also known for their yelping howls.

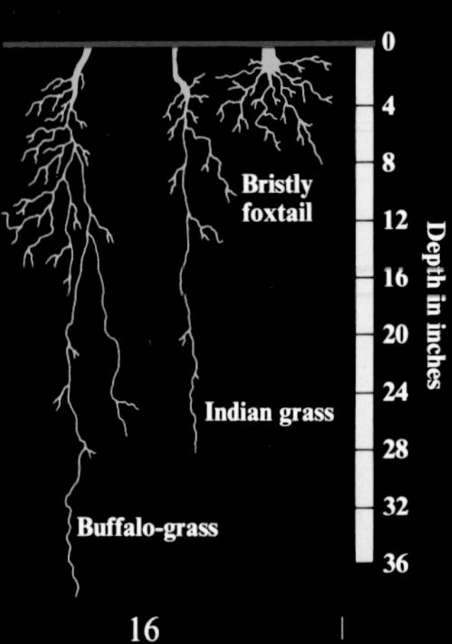

PRAIRIE WILDLIFE

One of the driving forces of any ecosystem is the interaction between predators and prey. In this respect, things have changed dramatically on the prairie over the last 200 years. The dominant predator of the prairie, the gray wolf, was exterminated by the beginning of the 20th century, and its largest prey, pronghorn sheep (11B) and bison (2C), nearly suffered a similar fate. Today, pronghorn populations are recovering: approximately 800,000 pronghorns now live in the western prairie and desert. And wolves may soon be reintroduced on the northern plains.

In the meantime, coyotes (9B) (which have actually extended their range since European settlement), eagles, and bobcats (lynxes) remain the prairie's most numerous predators, living on whitetail deer, cottontail rabbits, cotton rats, and other small game, together with scavenged meals and prairie fruits and plants.

With so much sky at their disposal, prairie birds offer some of the ecosystem's most spectacular shows. Wild turkeys are notable more for their size than for their aerial grace, whereas scissor-tailed flycatchers stage wheeling displays as they hunt flying insects in the autumn. Located beneath the North American continent's central flyway, the Great Plains are visited every year by millions of white pelican, Canada geese, and other waterbirds (19C) as they migrate between Canada and warmer climes.

dried or pounded into a mush, mixed with berries and nuts, and forced into a hide bag between alternating layers of buffalo fat, creating a *pemmican* that could last for years. Parts of the bison that the butchers would discard were invaluable; tendons became bowstrings and snowshoe webbing; shoulderblades became hoes; vertebral columns were covered with hide and turned into the skids of toboggans. Perhaps the bison's most valuable contribution was actually its dung: on the treeless prairie bison chips were the only source of fuel, burning hot and clean, and staying dry even in the rain.

Never again will the bison regain their pre-19th-century glory – yet they have staged a remarkable comeback in recent decades. More than 100,000 bison live on the prairie today, most of them in protected reserves or raised for their meat (which is higher in protein and much lower in fat and cholesterol than beef from cattle).

WALLOWS AND POTHOLES

Perhaps the most enduring reminders of the huge herds of bison that once dominated the ecosystem, "buffalo wallows" are found throughout the Great Plains. Shallow depressions up to maybe a couple of yards in diameter, the wallows (1D) were created by generation after generation of bison rolling in the mud to cool off and to discourage the attentions of mosquitoes on sweltering summer days.

Although no longer used by bison, wallows still fill up with rain, so providing food and drink for insects, birds, and small mammals. Prairie potholes resemble larger, deeper, more permanent wallows (19C). Although vital to migrating waterfowl, potholes have often proved to be simply a nuisance to humans. By draining and filling in the potholes and peatbogs in the prairies of Canada and the northern United States, farmers have contributed to a dramatic decline in the populations of duck and geese over the last 60 years.

PRAIRIE GRASSES

Because grasslands die every winter, continually returning their organic matter back to the soil, they boast some of the world's most fertile ground, with organic topsoil sometimes more than three feet deep.

Like a rain forest, a virgin prairie stores millions of years of evolutionary struggle and adaptation in the genes of its unique local plants. In the tallgrass prairie, the most productive grassland in the world, big bluestem (10D) grasses can grow to more than 9 feet high, yet their roots may extend 26 feet into the soil. The grasses above ground represent only 15 percent of the prairie's biomass. Other common tallgrass species include Indian grass and switch-grass (to the south) and fescue, needle- and wheat-grasses (to the north).

Typically, a tallgrass prairie has between 24 and 39 inches of rain a year. Westwards, this fades into mixed grass prairie where the precipitation drops to 14–24 inches a year. In the west, shortgrass prairie generally enjoys only 10–20 inches of rain a year. Although all these rainfall figures diminish northward, lower evaporation rates ensure that more water stays in the ground. The tallgrass prairie has a rich carpeting of native wildflowers, including Indian blanket (14D), prairie iron weed, and black-eyed susan (14D). Shortgrass prairie is dominated by buffalo-grass and blue grama grass, with little bluestem (1D) in sandier sites. Because water is scarcer where the shortgrasses grow, their root systems are more impressive even than those of tallgrass species – a single square yard of shortgrass sod can contain up to 5 miles of roots. Early settlers made use of these thick, dense, durable chunks of sod for houses. Once plowed, a prairie may take decades to regain its diversity. In western Colorado, where shortgrass prairies were plowed up half a century ago during an agricultural boom, plow lines are still visible from the air.

11 | 12 | 14 | 15

11 | 12 | 13

THE BISON

Before the arrival of European settlers, bison (2C) were the most successful large mammals on Earth. They shaped the prairies, whose grasses evolved their tough roots to withstand the attack of millions of bison hooves, which in their turn naturally aerated and tilled the soil. Because they ate mainly grasses, the bison gave shrubs a chance to compete, thereby maintaining the prairie's plant diversity. Once the herd moved on its continent-wide migration, the prairie would have months to rest. Fifteen to twenty million strong at the beginning of the 19th century, bison then fell victim to a massive slaughter. Hunters killed first for food and hides, then for tongues (between 1872 and 1873, 3,175 tons of buffalo tongues were shipped to the eastern United States and thence to Europe, where they were considered a delicacy), and finally for sport. Buffalo hunters like Thomas C. Nixon and Billy Dixon vied with each other to kill the greatest number of bison from a single stand: Dixon killed 101, but Nixon killed 120 – and they left the animals to rot where they fell. When the transcontinental railroad came through the prairie, passengers shot hapless bison from the windows. By 1895 only 800 bison survived in all of North America. Carcasses littered the plains in such quantities that early settlers earned a living by shipping bones back east to be turned into fertilizer, into carbon for sugar-refining, and to be mixed with clay to make china.

THE PLAINS INDIANS

Prior to the European invasion, the Native American People had also occasionally resorted to mass slaughter of the bison, for example by driving a herd over a cliff. But they generally made efficient use of the animals they killed. Fresh hides were stretched over wooden frames to make boats, or stitched into clothes using threads of bison leather. Meat was "jerked" into strips and

TEMPERATE GRASSLAND DYNAMICS

The key to survival

- Grazing, now mostly by domestic animals such as cattle, but originally and preferably by native species such as the bison
- Fire
- Low to moderate rainfall
- Undisturbed sod

Forces for change

- Plowing
- Overgrazing and undergrazing
- Topsoil loss
- Homogeneous planting of corn and wheat
- Insecticides
- Fire prevention
- Eradication of native species
- Invasion by exotic species, often introduced by man
- Drainage of wetland and potholes
- Highways, reservoirs, industry, and other commercial developments

In summer, whirling tornadoes often rip across the open prairie

Left to grow unimpeded, many prairie grasslands would eventually choke under the weight of their own dead cellulose. Yet for thousands of years, their growth was kept in check by the bison that once roamed in herds so large they sometimes took five days to pass a single point. Bison could strip a swathe of tallgrass down to its topsoil, and because they ate mainly grasses, they afforded forbs (herbage) and shrubs a chance to compete, maintaining the prairie's plant diversity. The bison's hooves, meanwhile, tilled and aerated the soil without destroying the prairie's tough roots. Once the herd moved on, the prairie had months to rest and recover.

But the bison was all but exterminated during the 19th century, and now it is beef cattle that graze the prairies. Properly managed, cattle can perform a role similar to the bison's in the prairie ecosystem, but suit it less perfectly. Because they have trouble digesting many prairie grasses, cattle graze mainly on legumes and other more delicate – and less competitive – plants. They require more water than bison, and tend to overgraze areas surrounding ponds. But the worst problem is that instead of migrating right across the continent, today's herds wander within the same fenced-off pastures all year round, never allowing the prairie time to recuperate fully. As a result, 88 million acres of land in North and Central America suffer from overgrazing and its attendant ills: soil erosion, loss of groundwater, reduction of plant diversity, and increased risk of flash flooding.

Ranching may have crippled the prairie ecosystem, but arable farming has utterly devastated it. Of the 252 million acres of prairie in North America before settlement by Europeans, only 3 percent has never been ploughed. According to the accounts of early settlers, ploughs made sounds "like zippers opening" as they cut into prairie root systems developed and strengthened over millennia. Broken up and turned upside down, the diverse, perennial prairie sod gave way to sweeping monocultures of grain-bearing annuals. Yearly plowing exposed the soil to wind and rain, causing one third of the prairie's topsoil to be lost to erosion. Insecticides, sprayed in ever-increasing amounts, worked their way up the food chain, killing off prairie birds and mammals, as well as their intended victims.

The prairie's shape and texture was once dictated by where rain fell, but farming has altered all this. Early farmers sank wells to overcome local water shortages. Then came widespread

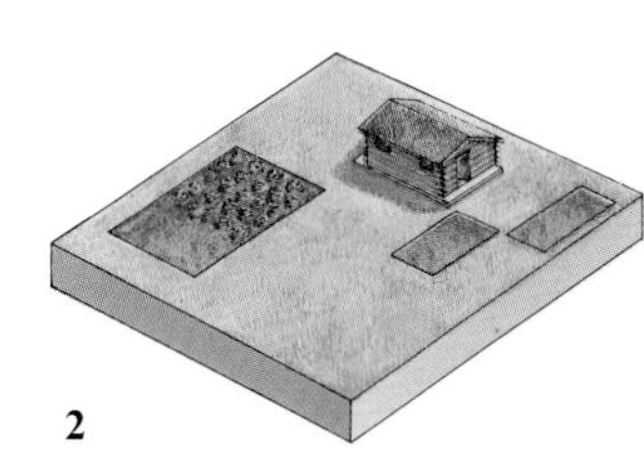

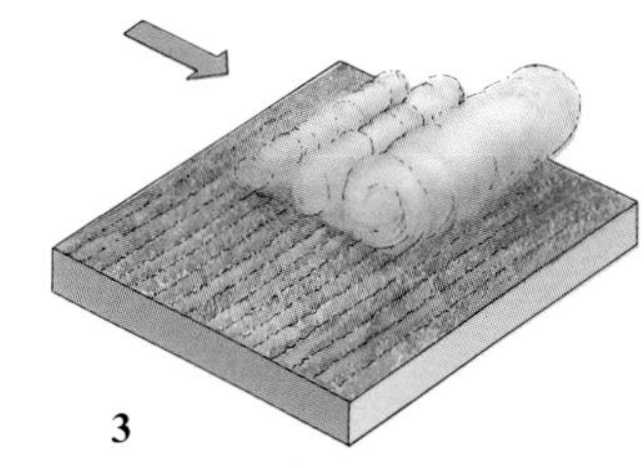

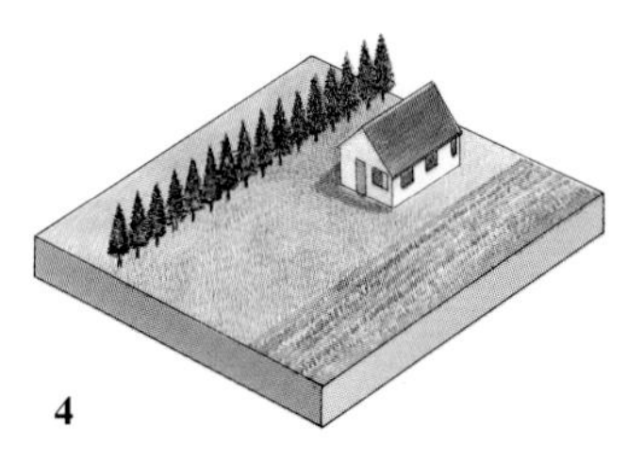

The North American dustbowl
Prairie plants spent thousands of years adapting to withstand the strong winds that buffet their ecosystem, along with the trampling hooves of bison (1). The prairie sod grew tough enough to fend off tornadoes, buffalo and drought, but yielded to the axes of American farmers, who used cubes of sod to build houses on the treeless plains and plowed up the prairie (2). As long as the rains kept up, crop yields were good. But when drought struck in the 1930s, the exposed topsoil dried to a fine dust. Soon prairie storms were stripping the soil, burying herds of cattle in dust. In 1934, one storm alone swept away 330 million tons of soil (3). The dustbowl left forced thousands of "Okies" (ruined farmers from Oklahoma) to pack up and head for California. Others stayed to restore the ecosystem they had helped to ravage. As part of the Great Plains Shelterbelt Project, more than 223 million trees in 18,550 miles of windbreaks were planted between 1935 and 1943. But the trees gave little shelter and drained groundwater from beneath the crops (4). The soil was finally saved by "trash farming" in which fields are only partly plowed, allowing old roots to hold down the soil. Even so, topsoil losses persist.

Soil exposed by crop farming is is quickly desiccated, then swept away by prairie storms

Lines of trees planted to shelter cropland from the wind have afforded only limited protection

irrigation. Arrays of sprinklers, pivoting around wells sunk deep into the Ogallala aquifer, create circles of cropland so broad that astronauts have spotted them from space. Genetically-engineered crops, never short of water, produce huge yields of grain. However, such productivity cannot be sustained indefinitely. At the present rate, modern farming will seriously deplete the prairie's topsoil and groundwater within the next century, leaving a desert behind. By contrast, the original prairie ecosystem survived more than 10,000 years, and showed no signs of weakening when the first farmers arrived.

The disruption is already evident among songbirds. Since the 1960s, American songbird populations have dropped by between 27 and 85 percent, depending on the region. Although much of this loss has been blamed on tropical deforestation, recent studies have shown that prairie residents are being hit as hard as tropical migrants. Imported species like starlings and sparrows have also taken over holes in which native woodpeckers and bluebirds once nested. Meanwhile, native species such as the brownheaded cowbird have taken advantage of the prairie's changing character to expand their numbers: they are brood parasites, leaving their eggs in other birds' nests. Their chicks hatch quickly, absorbing all of their surrogate parents' attention. Thanks to overgrazing, cowbirds have been better able to spy on other nesting birds and find foster homes for their eggs. In so doing, they have helped to decimate populations of orioles and black-capped vireos. Clever, rapacious, and opportunistic, cowbirds may be appropriate victors in the latest, most violent chapter in the prairie's evolution.

earlier in the year.

Forests in temperate Europe have a long spring, the woodland herbs becoming active as early as January and showing a gradual development over the following four months. Continental forests in Europe and, most notably, in North America, show a much more rapid development of the herb stratum as the snow melts in late March, April or sometimes in early May, an event which is then quickly followed by the trees coming into leaf.

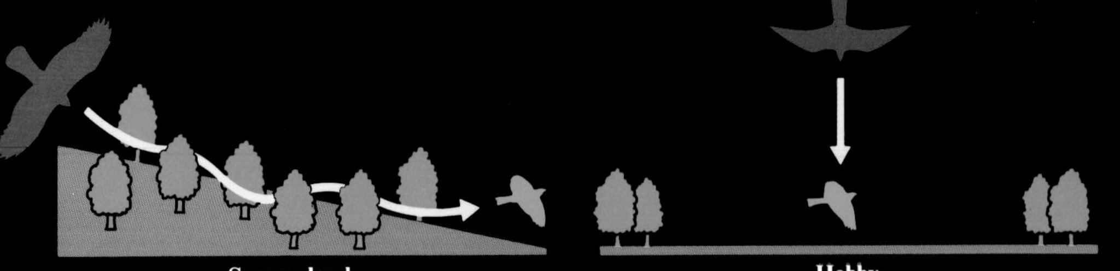

SPEED AND AGILITY

The rounded wings of the sparrowhawk (14A) help make it very maneuverable through the woodland canopy as it swoops along a winding flight-path, then strikes lightning-fast at small hedge-birds. The hobby – another bird of prey that frequents the New Forest in summer – is designed for high speed but lacks maneuverability. It hunts fast-flying insects and birds, such as swallows, in open situations.

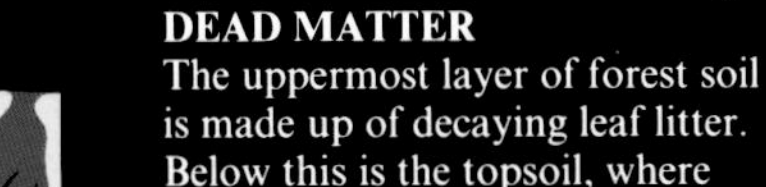

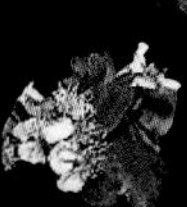

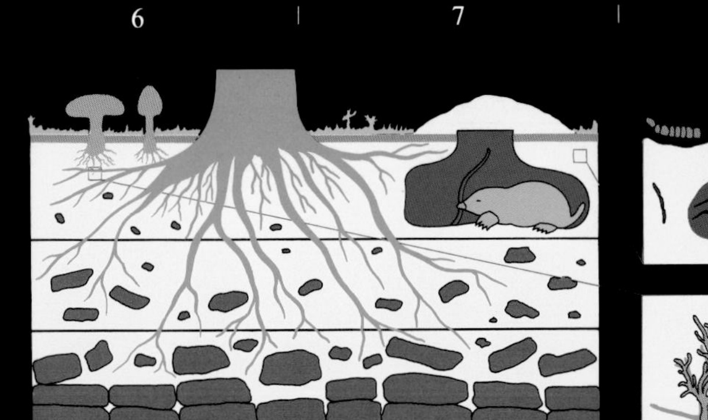

Main image: soil with roots, fungi, mole, and earthworm
First inset: small detritivores – springtail, nematode worm, red mite
Second inset: the coral-like growth of mycorrhizae

DEAD MATTER

The uppermost layer of forest soil is made up of decaying leaf litter. Below this is the topsoil, where the organic matter of the litter has decomposed to form humus. Because in a deciduous forest the major leaf-fall occurs in autumn, as temperatures are falling, there is little decomposer activity and the litter forms a blanket on the ground, which can shelter hibernating animals such as hedgehogs. In spring the decomposers multiply and become active in the rising temperatures. Earthworms and woodlice eat the leaves and produce feces that are themselves nutrient-rich, on which springtails and smaller worms feed. Their feces are in turn eaten by mites, whose own feces are broken down by protozoa.

The soluble nutrients liberated by this process are washed into the soil to be reabsorbed by plant roots. Mycorrhizae are symbiotic relationships between plant roots

Little of the primeval temperate forest now remains, and nearly all surviving fragments have been greatly affected by humans and their animals. Some areas nonetheless retain much of the diversity of the original forest ecosystem, especially the ancient medieval hunting forests that have been preserved up to the present day. The woodlands of the New Forest, in England, have enjoyed a continuous history of forest cover from the original wildwood, and many of the animals and plants of mature lowland temperate forest still live there.

SPRING FLOWERS

One of the most striking features of the New Forest is the abundance of spring flowers such as anemones and bluebells (1D). This abundance occurs because the herbs have the advantage over the shrubs and trees in being able to come into leaf much earlier, undertaking most of their active photosynthesis and growth before the spreading leaves of the trees and shrubs make the woodland dark and uninviting.

The rising ground temperature regulates this window of opportunity. It is easy to underestimate how different the temperature may be in different parts of a wood. On a sunny day in early May, the air temperature just above the herbs on the ground may be 63°F warmer than air that is only 12 inches higher up. The activity of herbs, shrubs, and trees, and their progressive leafing out – their phenology – is tied to this graduated warming of the forest. Close to the ground the temperatures that promote active plant growth are reached much

TREES

Most of the biomass in temperate forests resides in the woody tissue of trees and shrubs, or lives on decaying wood. The height and complexity of the trees, the diversity of the tree and shrub species, and their response to varying soil conditions and water availability, make forests the most structurally complex and biologically diverse of all habitats.

The tree species that make up the temperate forests differ greatly in their ecological requirements. It is a mistake to think that all trees are broadly similar simply because they may be similar in appearance. Many trees are as evolutionarily dissimilar to one another as humans are to fish. There are consequently many different strategies among temperate forest trees in connection with reproduction, tolerance of shade, reaction to damage and disease, and the ability to grow on soils of different sorts.

The trees of the New Forest provide some excellent examples of this wide range of response. Birch trees are short-lived pioneers, living for perhaps a century but producing millions of tiny well-dispersed seeds during that time. Oaks are long-lived pioneers: once established, they persist for several centuries. They produce fewer seeds each year, but the seeds are well stocked with nutrients and are widely dispersed by forest animals and birds, notably jays (15C). The price paid for this dispersal is that most seeds are consumed. Oaks

(5A) are very susceptible to shade, and thrive only if established in open, well-lit conditions. Beeches (12B), by comparison, are very tolerant of shade and can flourish in darker areas of established woodland among such pioneer species as birch or oak. Beech persists for several hundred years in favourable conditions, but is prone to natural catastrophes, succumbing in great numbers to drought, storm, or disease.

TEMPERATE FORESTS

No more than six thousand years ago vast areas of eastern North America, Europe, Asia, and the Far East were blanketed in dense, unending forests of deciduous and coniferous trees – oak, beech, hickory, birch, aspen, and pine. Wherever there was enough water, temperate woodlands like these grew thick and lush, providing a home for all kinds of plants and animals. However, no ecosystem has been quite so savaged by human activity: temperate forests have been reduced to a tiny fraction of their former extent, and cities and farmland now swarm over the areas where woods once grew. Virtually no virgin deciduous forests survive in Europe and only isolated pockets are still to be found elsewhere in the northern hemisphere. Even where natural temperate forests do grow, they are often very different from the primeval pristine forest of long ago and betray some traces of human management or the effects of human activity. Only in the southern hemisphere, in places such as southern Chile and western New Zealand, is the natural forest relatively untainted. Here, though, conditions are so wet that the forest is sometimes classified as temperate rain forest.

In their natural state, temperate forests are changing all the time, as old trees are felled and young saplings grow through, creating a complex mosaic of ancient and new woodland. Every now and then, storm, flood, disease, drought, landslip, and even fire cuts a swathe through the wood, leaving the way clear for young trees to emerge. Some stands remain undisturbed for several centuries, and here mature and ancient trees grow dense. However, on steep slopes or near flood-prone rivers, catastrophes come frequently and dramatically, and few trees survive to reach maturity. So the character of each area of the forest depends to a large extent on just how often and how severely natural disasters strike – that is, its "disturbance regime."

Even within the individual stands there is tremendous variety. Here and there, dense wood is broken by broad sunny glades or small gaps where trees have fallen; saplings, shrubs, and woodland flowers grow here in profusion. In places, saplings are destroyed by browsing animals such as bison, aurochs, and, in eastern Asia, wild cattle, leaving large areas as scrub or even open grassland. Where water is present, beaver ponds and meadows, oxbow lakes and open riverside mud and shingle all provide a wealth of habitats for a rich variety of plant and animal life.

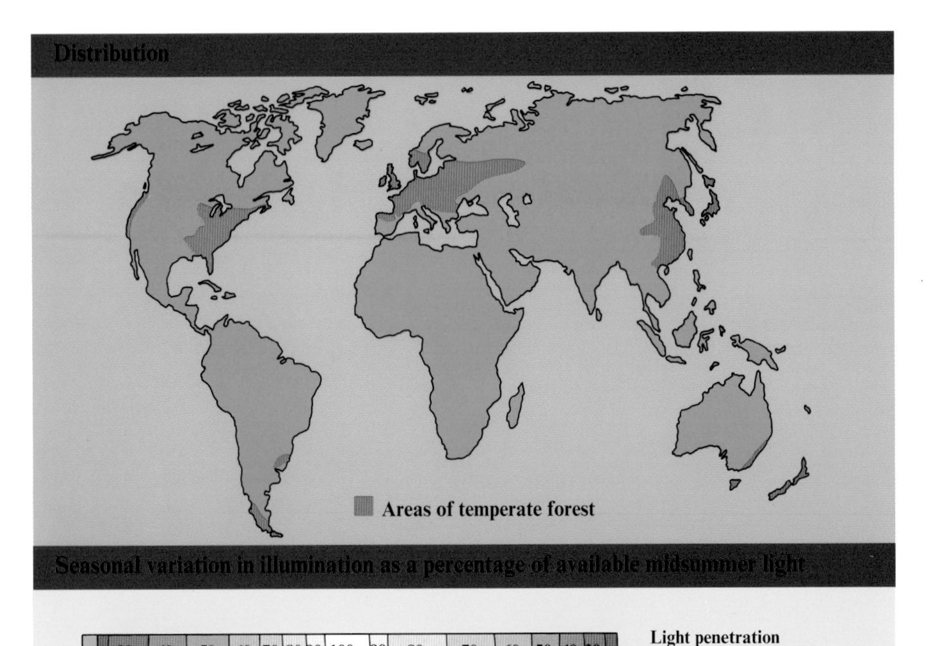

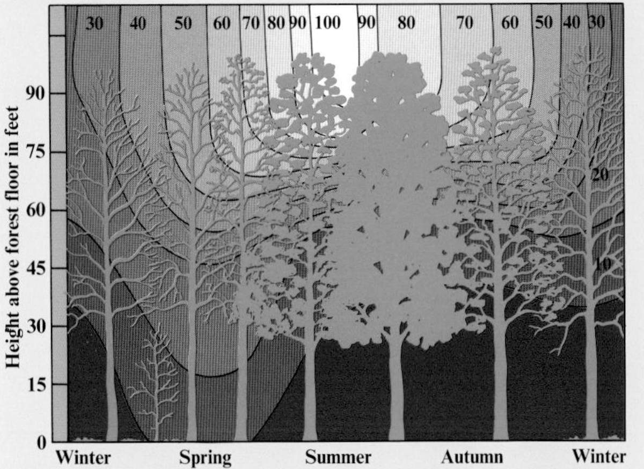

Light penetration
Although there is most sunlight in midsummer, the leaves of a temperate forest prevent more than about 10 percent of this from reaching the forest floor. In fact, the ground-level illumination in midsummer is less than it is in winter. Most light reaches the forest floor in spring, before the leaves have begun to open, and promoting the characteristic sudden profusion of spring flowers.

Forest trees

In most temperate forests in eastern North America and western Europe, the canopy is formed by broadleaved deciduous trees such as oaks (*Quercus* spp.), beeches (*Fagus* spp.), maples (*Acer* spp.), ashes (*Fraxinus* spp.), and limes (*Tilia* spp.). But where winters are colder, nearer the poles and farther from the sea, deciduous trees are mixed with or even replaced by conifers such as pine, redwood, hemlock, and cedar. In very wet areas, such as western New Zealand, Tasmania, Chile, and British Columbia, the forest is classified as temperate rain forest and is dominated either by broadleaved evergreens or conifers and is filled with ferns and mosses.

At their best, temperate forests contain dozens of different tree species, but in European forests the variety was significantly reduced during the Ice Ages. In any one area of temperate forest there are usually just a few dominant species: in western Europe, for example, beeches, ashes, or limes tend to grow on limestone, while oaks, birches, and hazels flourish on clays or more free-draining gravelly soils. Sometimes, however, there is a broad mixture of trees present. Each type of woodland – whether beech, oak, or mixed – fosters its own woodland plant communities, and any one tract of forest is likely to contain at least three or four such communities, reflecting soil conditions and rainfall.

Some 18 sq miles of Poland's Bialowieza National Park are set aside as "strict reserve." This is some of the last primeval forest in Europe

Most existing European woodland has at least some history of management, having been used to provide timber and undergrowth

Summers in temperate forests are often warm, but winters can be bitter and ground temperatures usually drop below 41°F. When the soil gets this cold trees cannot obtain water properly and suffer virtual drought. This is why so many temperate forest trees are deciduous, shedding their leaves in winter. Evergreen species that retain their leaves, such as hollies (*Ilex* spp.), yews (*Taxus* spp.), and ivies (*Hedera* spp.), adapt to drought by means of waxy leaves and fewer exposed stomata.

Yet the winter leafdrop, although it stops trees growing, lets sunlight through to plants below the canopy, encouraging a rich understory of shrubs, herbs, ferns, and mosses. The forest floor changes from season to season as the amount of sunlight penetrating the trees varies. In summer, the leaves grow thick and only a few flecks of light reach the ground, allowing only shade-tolerant moisture-loving plants such as wood sorrel to thrive. In winter, the trees are bare, but light levels are too low for anything but ferns, mosses, and a few herbs to grow. In spring, however, days brighten before the leaf canopy closes over, and the forest floor bursts into life. At this time of year, many deciduous woods are carpeted with banks of bluebells, scillas, anenomes, and other woodland flowers.

The lives of forest creatures vary with the seasons too. Food supplies are most abundant in late summer when the trees are heavy with nuts, and apples, blackcurrants, and hawthorn fruit briefly. At this time of year the forest is teeming with life. However, with the coming of winter activity dwindles as nights grow cold and food becomes scarce. Many woodland birds migrate to warmer places, while small mammals, such as hedgehogs and chipmunks, hibernate, conserving energy by sleeping right through the long winter months.

In summer, when the leaves are on the trees and days are warm, the canopy is full of birds such as warblers and finches. There are squirrels, too, but most woodland mammals are ground-dwellers – badgers, bears, red deer, wild pigs, moles, and rodents, preyed upon by wolves, wild cats, and foxes. Many of these species have suffered considerably from the fragmentation of forests. Small areas of woods are not enough to support the larger predators, and creatures such as the brown bear in Spain and the wolf in Scandinavia have been pushed to the verge of extinction.

Deadwood homes

All natural temperate forests are filled with an enormous quantity of deadwood in all stages of decay, from fallen twigs to tree trunks that are dead but still standing. There is deadwood too within the heart of living trees, and in dead branches still attached to the crown. Some deadwood is waterlogged and wet, such as rotting roots in wet soil or branches fallen onto wet ground; some is very dry, and exposed to extremes of heat and cold. Perhaps 40 percent of the forest's entire biomass is deadwood, with another 30 percent as living wood.

All this deadwood offers a wealth of potential habitats for plant and animal life. Each kind of deadwood provides a home for a particular range of species, and this is why untouched temperate forests hold such a diversity of insect and fungal life. Where humans remove deadwood for fuel and timber – as they have done so widely in Europe and North America – they destroy the habitats of these plants and animals with the result that deadwood-dwellers are now among the rarest of woodland species. Beetles such as *Rhysodes sulcatus*, once common in the plentiful deadwood of the primeval forest right across Europe, now survive only in a few isolated pockets in the Alps and the Pyrenees. Equally rare are once-abundant bird species such as the white-backed woodpecker in western Europe, and the ivory-billed and red-cockaded woodpeckers of North America.

The primeval forest of Bialowieza is home to many creatures extinct elsewhere in Europe

exceptionally well developed. On the underside of the thoraxes of ambrosia beetles, for example, is a cup that scoops up the spores of the fungi on which they depend, thus ensuring the dispersal of the fungi to new deadwood sites, and ensuring a continuing food supply for the beetles. Other types of fungi have sticky spores that adhere to the bodies of wood-boring beetles in order to effect similar dispersal.

Deadwood is low in nutrients, comprised as it is principally of carbon together with minor quantities of nitrogen and phosphorus. One result of the very low food content – and coincidentally of the comparative lack of danger in deadwood habitats deep inside rotting trees – is that wood-boring beetles and large flies take a long time to mature. The giant stag beetle and the longhorn beetle (7D) may take four to five years to reach full size. This long maturation, however, does mean that they can grow very large indeed.

REGENERATION

Whereas beech trees (12B) are sorely afflicted by catastrophe, many other species – lime and hazel, for example – sprout vigorously when broken or felled: lime from broken branches, stumps, or roots; hazel from an underground stool that throws up a dense clump of shoots. Some species do not sprout when the main stem is cut or felled, or becomes diseased, but send up many hundreds of young suckers: cherry and aspen trees are good examples of this. Elms send up suckers and can use them to invade established stands of other species such as hazel or ash, provided that soil conditions are favorable – elms require nutrient-rich soils and cannot compete with oaks or beeches on drier, less fertile soils, or with alders and similar species on the wettest, waterlogged soils.

The ability of alder to put up with wet soil and to fix nitrogen-using bacteria in nodules within its roots affords the tree a substantial advantage over its competitors in moist conditions.

The complex dynamic interactions between the trees and shrubs as they respond to changes in soil and water, and to catastrophic changes such as storm or drought, serve to maintain the species diversity of trees in the temperate forest. Although there may be a few dominant species, this diversity contributes to the great number and variety of woodland communities existing within the temperate forest ecosystem.

ACTIVE WINTERS

Those animals that do not migrate like the woodpecker (6C) or hibernate like the squirrel must forage to supplement any fat reserves they might have managed to build up during the autumn. Deer (11C) grub for roots or browse on bark and twigs. Some birds, such as the tits, hunt dormant insects and their eggs, while the overwintering woodcock (18D) searches under the snow for insects and earthworms.

GRAZERS AND BROWSERS

For a thousand years, the principal animals to graze in the New Forest have been the king's deer and the commoners' livestock. Red, roe and fallow deer (11C) can still be found there. The grazing and browsing of cattle and ponies (7C) have a great effect: keeping the forest floor open and well-lit, and promoting tree species such as beech, oak, and holly that are resistant to browsing when young, while eating back less tough and more palatable species such as ash and hazel. Because of its palatability, the native lime tree disappeared from the forest many centuries ago.

Kept down by constant grazing, stocks of young trees are replenished only at irregular intervals when for some reason the grazing pressure drops, and the forest trees tend to correspond to discrete "generations." One generation can be dated to the Deer Removal Act of 1851, when deer were removed from the forest, allowing a surge of regeneration.

The forest clearings with their sunny, bramble-strewn edges support many birds and insects such as the chaffinch (10D) and speckled wood butterfly (8C).

DEADWOOD BEETLES

Of the deadwood, or "saproxylic" beetles, some feed only on wood that has completely dried out; others require waterlogged or at least very wet wood; yet others prey on the larvae of beetles that bore into the rotting heartwood of trees. The larvae of the *Staphylinidae* are adept predators, hunting other beetles through powdery, dry wood dust.

The co-evolutionary relationship between some beetles and fungi is

11 | 12 | 13 | 15

12

and fungi that for many plants are essential to this process of reabsorption. Mycorrhizae occur when fungal filaments penetrate the cells of a root at the same time as spreading out into a network in the soil, functioning like extended root hairs. They help in the decomposition of litter, and in the absorption of nutrients. At the same time they form a barrier to harmful bacteria and stimulate the plant into producing anti-bacterial chemicals. Some trees, such as oaks and birches, could not survive without mycorrhizae. For their part, the fungi depend upon the tree roots for an energy source and cannot germinate until they find a root. The above-ground fruiting bodies of mycorrhizae are the plentiful and varied mushrooms (13D). Some types of mycorrhizae have underground fruiting bodies instead: the truffles, which rely on mammals such as voles to spread their spores.

A huge amount of the available biomass in temperate forests occurs as dead and decaying wood. In an undisturbed forest – one that has not been modified for hundreds of years by logging or by catastrophe such as a hurricane – up to 40 percent of the biomass may occur as deadwood of various dimensions.

Few organisms are capable of exploiting the energy and nutrients locked up in deadwood. In fact, only fungi and the protozoans within the guts of termites are able to break down the complex cellulose and lignin molecules. The most obvious of these in the forest are the bracket fungi (4B) and the fruiting bodies of fungi that consume the heartwood of trees.

Despite its indigestibility, many creatures make use of deadwood, mostly by eating the fungi that can break it down or by eating the byproducts of that breakdown process.

TEMPERATE FOREST DYNAMICS

The key to survival

- Neglect of fertile land in a region of moderate, seasonal climate will naturally promote the growth of woodland
- Many woods have a long history of management, which must be maintained to preserve the current species

Forces for change

- Logging for timber
- Acid rain
- Pollution from airborne nitrates
- Toxic metals such as aluminum being absorbed as ions by tree roots
- Replanting with commercial, but ecologically unsuitable, trees
- Clearing the forest to provide pasture and agricultural land
- A cessation in the pattern of cutting and coppicing that historically has done as much to shape the ecosystem as any environmental factors
- High winds causing tree falls
- The introduction of inappropriate species, or the extinction of existing large browsers

Very little temperate forest remains completely untouched by human activity, except perhaps in the Ussuriland forests of Siberia and in northern China. All across Europe and North America, temperate forests have been cleared for agricultural land or pasturage, and the areas of woodland that remain have been modified a great deal by woodland management for food and timber over past years.

The ecology of the woods that survive reflects their past management. Those managed as coppices in which the wood is cut over a very short 10–30-year cycle are often rich in the plants and butterflies of the glades and woodland edges, but have almost completely lost the beetles, flowers, fungi, and lichens that depend on old or ancient trees, decaying wood and mature bark. The much rarer "wood-pastures" – the commons and forests where grazing livestock and old-growth trees have been kept together – often have many species of insects and plants associated with old trees. But they do not have the flowers and insects of more natural woods, nor the understory of hazel, elm, hawthorn, and lime that browsers find so palatable.

Because certain soils are better for farming, such as those in river valleys, woodland clearance has been by no means even. Some kinds of wood have been especially hard-hit. In Europe, most surviving forests are on less fertile, acidic, mountain soils a far remove from the natural habitat of certain trees such as the black poplar and various elm species in western Europe. This is why these trees are characteristic of farmland landscapes.

Sometimes, cleared woodland is left to regenerate. But the new, secondary wood is very different from that of older, well established woodlands. Most temperate wood is secondary, which is why native woodlands are so highly valued by ecologists. Even many very ancient woods are secondary in origin, and still lack some characteristic woodland plants despite many centuries of existence. Some woodland plants are very poor colonists and their presence indicates that the woodland is of ancient origin. In Britain, lime, herb paris, and wood anemone are all known to be poor colonizers of new woodland, and are thus regarded as prime indicators of age. Elsewhere in western Europe, other species – such as ivy in Poland – may prove to be better evidence of long woodland continuity.

Effects of glacial ice advance in Europe and North America

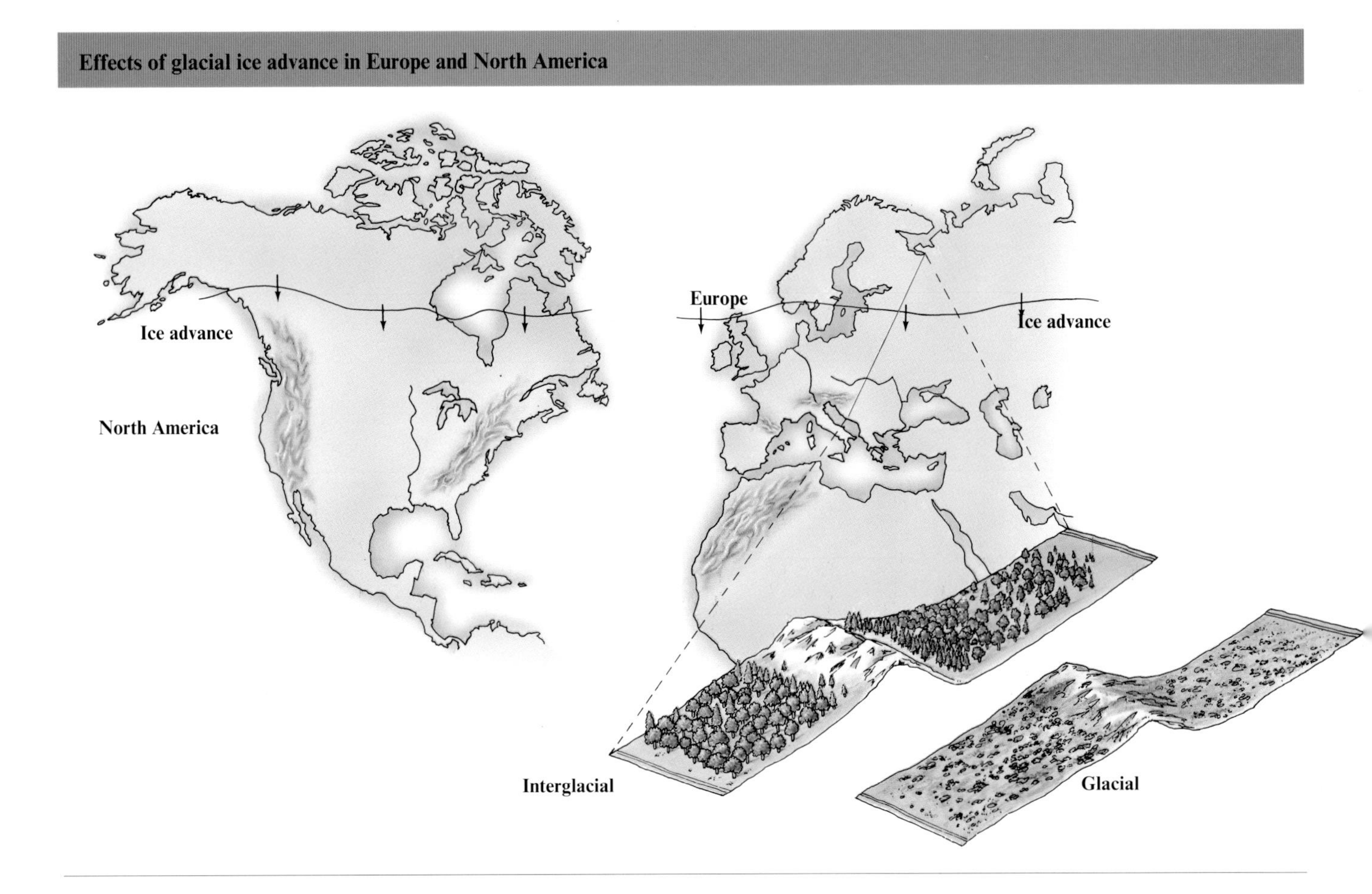

Trapped species
The temperate forests of North America contain five times as many tree species as the forests of Europe. They also contain many more amphibians, for example, than the frogs and the few newts (above) of Europe. This is essentially because more species were extinguished during the Ice Ages in Europe than in North America – not because conditions were more severe, but because of the physical geography of each area. In Europe, the major mountain ranges, such as the Alps and the Pyrenees, are aligned east-west and presented a major barrier to species migrating south in front of the advancing ice sheets. In North America, mountain ranges such as the Rockies and Appalachians are aligned north-south, allowing species to move south, then return when the ice sheets retreated. Just how much of a barrier mountains present to the spread of species is indicated by the amazing diversity of salamanders in the southern Appalachians. Salamanders find mountains more of a barrier to dispersal than most creatures, and the rugged terrain here allowed many different sub-species of *Plethodon jordani* salamanders to evolve in isolated valleys. "Geographic isolates" like these may well be on the way to becoming separate species.

Forest management
Even where forests have never been cleared altogether, as is often true in North America, they have usually been altered in some way – by logging, for example. Logging radically changes the species composition of a forest by promoting a prevalence of light-loving pioneer species over large areas, and alters the age structure of the forest toward a predominance of younger trees. Large-trunked, big-branched deadwood in damp shady conditions is gradually replaced by lesser amounts of small-limbed deadwood in hot, open, sunny areas. This in turn leads on to a dearth of deadwood habitats that may continue for several centuries, even if the forest remains untouched after its initial logging. Species dependent on deadwood and mature trees (many insects, lichens, mosses, and fungi, in addition to specialized mammals and birds, such as flying squirrels and old-growth-dependent woodpeckers) are consequently extremely rare across these forest ranges.

In western Europe there has been a long tradition of managing woodland, dating back to at least the early Middle Ages. The woods were managed as coppices. Trees such as hazel and ash were cut repeatedly in 5–35-year cycles to produce coppice "stools." Interspersed among them were a few standard trees, usually oak, grown on 50- to 100-year cycles to produce timber. The coppiced hazel and ash were used for anything from hurdles to spears, pikes and arrows. The timber trees were used in the construction of houses, bridges, and other works. But the wood's primary value – until the arrival of cheap coal and oil – was as a source of firewood for fuel and for making charcoal.

The long history of intensive management of such coppice woods has left them devoid of most species associated with mature trees, but very rich indeed in the plants and animals – especially butterflies – of early woodland succession: of open spaces, glades, and forest edges. Their continued management is consequently a very high priority for nature conservation if these elements of the woodland fauna and flora are to survive.

More subtle changes have been wrought over very long periods by the removal of large game, such as aurochs (extinct in Europe by the 17th century), or of predators such as wolves and bears, which has the effect of increasing the numbers of grazers. Either a removal of animals or an increase in their numbers can radically change the browsing pressure on regenerating trees in the forest. In recent centuries, the grazing of livestock and game (in Royal Hunting Forests, for example) may have greatly increased the abundance of browse-resistant species such as beech or oak, while decimating more palatable species such as ash, hazel or lime.

Threatened pockets
Pristine or near-pristine old-growth forests are now to be found only in very small pockets (of no more than tens or thousands of acres) in Europe and North America, and these are often set in much larger tracts of disturbed or managed forest.

The largest of these is the vast Bialowieza forest on the Poland-Belarus border. Much larger tracts of genuinely old growth still remain in the Ussuriland forests of eastern Siberia, but even these have recently come under threat from logging as a result of the opening up and development of the former Soviet Union.

Bruceid beetles lay their eggs on the spiral seed pods (right). These hatch and the larvae feed on the seeds, killing them (A). However, if a grazing animal eats the seed pod before any damage is done, the seeds pass through its digestive tract unharmed but the eggs or larvae are killed (C). Browsed, beetle-infected seed pods are even more successful than unbrowsed, uninfected pods (B), as the seedlings are helped by moisture and plant food in the animal dung.

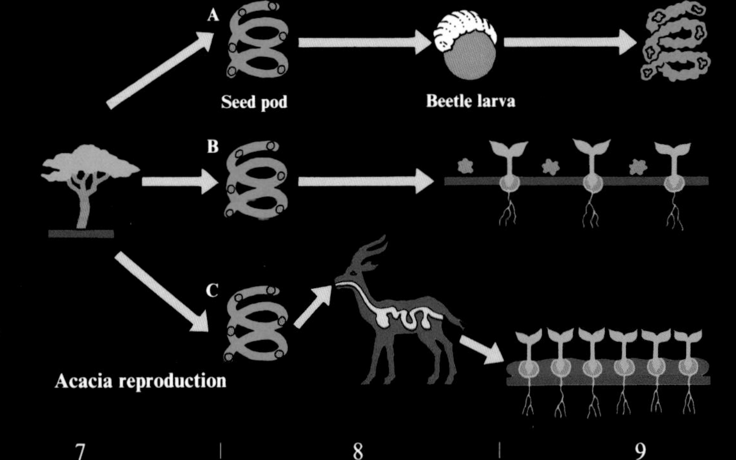

SCAVENGERS
With its huge wingspan (up to 10 feet) the lappet-faced vulture can soar effortlessly on rising air currents, traveling huge distances in search of animal carcasses. The cornea of its eye has a special magnifying region which allows it to spot signs of feeding from a great height. Vultures are voracious feeders, their long necks and bare heads perfectly adapted to exploring inside a rotting carcass. They never kill big game, but with other species the relationship between predators

6 | 7 | 8 | 9 | 10

6 | 7 | 10

allow much greater movement of its limbs than would otherwise be possible. When changing direction at high speed, the cheetah uses its tail as a stabilizing rudder. Because it uses so much energy to achieve its speed, the cheetah has little stamina, and must catch its prey within a few hundred yards, or give up the chase. In the past, the cheetah's speed was harnessed by man. The Egyptians used cheetahs for coursing game, while in 16th-century India the Emperor Akbar kept a thousand trained cheetahs for hunting.

Savanna covers much of Africa, from the subdesert that fringes the Sahara and Kalahari deserts to the vast grasslands of the East African plains. The plains play a particularly important ecological role, supporting large numbers of wild mammals as well as providing grazing for more than 70 percent of the region's livestock. During the wet season they turn green, as grasses and trees come to life in response to the equatorial rains. As the dry season progresses, the savanna takes on a parched, golden appearance.

TREES
The greatest threats to savanna trees such as the baobab (14B) and acacia (4B) are their periodic destruction by fire and damage by browsers like the giraffe (3C), impala, gerenuk, and elephant (12C). The flat-topped acacia trees tend to be small (6–20 feet high) and short-lived (up to 40 years). Their leathery leaves and defensive thorns help to deter herbivores, and some protection from fire is afforded by their thick bark, which often contains fire-retardant substances. While the baobab's bulbous trunk consists largely of water-storing cells, the acacia has no water reserves to sustain it in time of drought. For this reason, some species of acacia thrive only on dried-out riverbeds, where their deep roots enable them to tap into groundwater reserves that sustain them through the dry season. Although for most trees browsing is destructive, the reproduction of one of the common acacias (*Acacia tortillis*) is helped by the presence of large grazing animals.

PREDATORS AND PREY
The savanna is an open landscape where prey cannot readily hide from predators and speed is often the only escape from being eaten, or the only guarantee of a meal. The East African plains are, not surprisingly, home to some of the world's fastest land animals, including the cheetah (8D) and its prey – gazelle (7D), and other small antelope. Other hunters, including wild dogs (7C) and lions (9C), use cooperative behaviour to make their catch, which allows them to hunt larger antelope such as wildebeest (11C), and even to tackle prey as large as wild buffalo if they are desperately hungry. Scavengers, including hyenas and vultures (13D), are always on hand to take whatever is left over, or even to rob an exhausted cheetah of its newly caught prey. The leopard (18B) avoids this pitfall by dragging its kill to safety high in the branches of a nearby tree.

HIGH-SPEED CHASE
The cheetah uses an explosive burst of speed, sometimes accelerating up to 55 mph in under five seconds, to catch its prey. To achieve this, it completely leaves the ground twice during its running cycle, once with its legs fully extended and once with them bunched under its body. During each of these short "flights" the cheetah can cover an incredible 12 feet. All this is made possible by the cheetah's flexible backbone and the way in which its shoulder blades and pelvic girdle swivel to

SAVANNA

Widely distributed throughout the tropics, the vast open spaces of the world's savannas are among the most spectacular of all landscapes. These huge expanses of grass, interrupted only by scattered trees, are home to massive herds of grazing animals as well as some of the most awe-inspiring of the world's predators – the lion, leopard, and cheetah.

The term savanna comes from the 16th-century Spanish word *zavanna*, meaning "treeless plain." Today it is used to describe a more varied set of landscapes – from the truly open South American grasslands to the light woodland of northern Australia – that share particular biological and physical characteristics. Savanna occurs where rainfall is too low to support rain forest but sufficient to ward off desert. It is held in a delicate and highly productive balance between these two extremes not only by climate but also through the effects of grazing and burning.

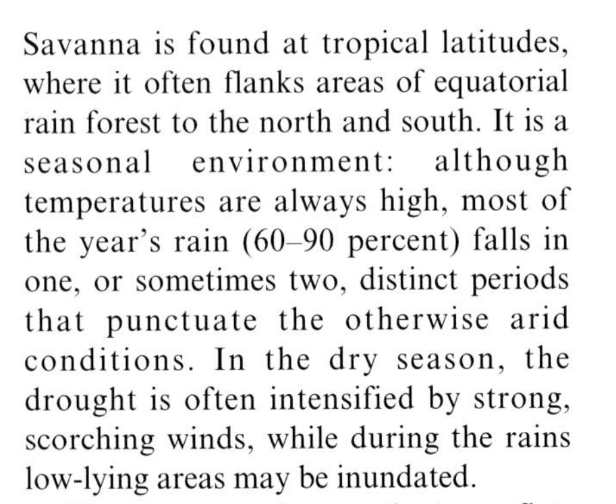

Savanna is found at tropical latitudes, where it often flanks areas of equatorial rain forest to the north and south. It is a seasonal environment: although temperatures are always high, most of the year's rain (60–90 percent) falls in one, or sometimes two, distinct periods that punctuate the otherwise arid conditions. In the dry season, the drought is often intensified by strong, scorching winds, while during the rains low-lying areas may be inundated.

The most conspicuous plants are flat-leaved grasses, which may be up to 5 feet in height. Typically, one or two species dominate a given area, forming a more or less continuous carpet which all but smothers the growth of small herbs. Trees and shrubs are often present, but trees are usually sparse. Many are deciduous, losing their small leaves in the height of the drought in order to conserve water. Only after the rains does the savanna come alive with an explosion of annual species.

Energy is cycled quickly in the savanna ecosystem. Little is locked up in long-lived structures, such as trees, and grasses grow prodigiously in the wet season, recovering swiftly after being grazed. Savanna produces 4.5 lb of plant matter per square yard of soil each year, half the amount produced by rain forest. This level of productivity, however, is attained with a tenth of the biomass of rain forest, making savanna one of the most efficient of all terrestrial ecosystems. It can therefore support a rich variety of grass- and seed-eating insects, birds, and mammals: in East Africa, more than 20 species of large, hoofed herbivore can be found in some regions. However, some of these species are only temporary inhabitants of any particular region, moving from place to place as they follow the rains, which are highly seasonal. Australian savanna, while poorer in species, has huge numbers of marsupial grazers – the kangaroos and wallabies. The rabbit, an introduced species, has also adapted to Australian savanna conditions, and at one time the numbers of rabbits approached epidemic proportions. Central and South American savanna, however, can lay claim only to a variety of grazing rodents, two species of rhea and two species of deer.

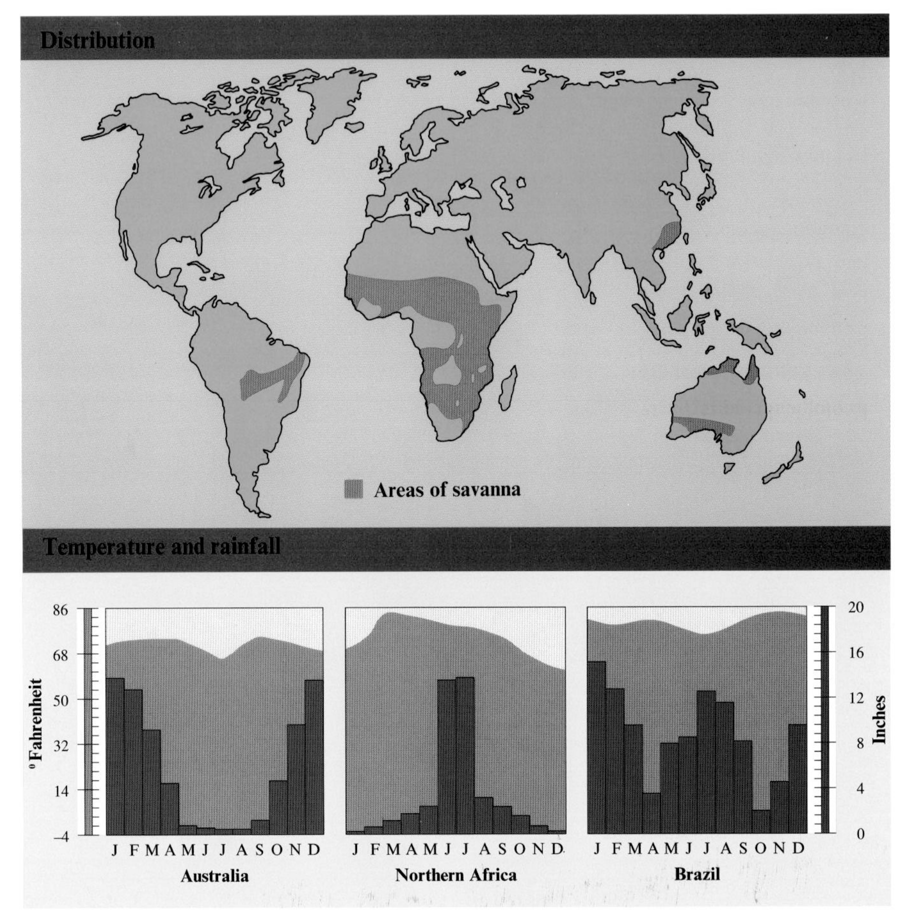

Burning

Grass fires are a frequent occurrence in the dry season. While they may appear to be destructive, fires are in fact essential in maintaining the ecosystem: without them the savanna would soon be lost, covered over by dense woodland. Burning destroys the tender shoots of young saplings. Because trees are slow-growing, only a few individuals avoid the fires for long enough to achieve a height that puts their leaves out of reach of the flames. As a consequence, grasses, which are fast-growing and regenerate quickly after burning, thrive at the expense of the trees. The effect of fire is greatly reinforced by the activities of herbivores, which prefer the protein-rich foliage of trees to the dry, fibrous leaves of the typical savanna grasses.

Human activity has been, and still is, important in making and maintaining savanna. While some fires on the savanna are caused by lightning, most are started deliberately. Controlled burning by indigenous peoples creates nutrient-rich ash, which mixes with the soil, encouraging fresh growth of the grasses on which their cattle depend. Ancient savanna peoples also set fires, using the advancing flames to drive animals toward their hunting parties, thereby saving hours of tracking and stalking. Many animal species, notably African storks and secretary birds, use a similar tactic, haunting the edges of natural or controlled fires and picking off small insects and rodents as they scatter in panic before the flames.

Human origins

The boundary between savanna and rain forest is an evolutionary forcing-ground: rain forest species that colonize the savanna (or vice versa) must adapt rapidly to a very different environment. Under such conditions, new species are soon formed, and many anthropologists believe that the early stages of human evolution took place in the African Rift Valley, where five-million-year-old Hominid fossils have been discovered.

Ecological counterparts

The African ostrich, Australian emu, and South American rhea fill the same ecological niche in their savanna ecosystems. These large, flightless birds all form grazing flocks and rely on long, powerful legs and high-speed flight to escape predators. The ostrich is the largest surviving bird, sometimes reaching 8 feet in height and 375 lb in weight. The emu is somewhat smaller, at a height of 6 feet, while the rhea rarely grows any taller than 5 feet. Though similar in appearance, these birds are only distantly related: it is possible that their common ancestors lived in the ancient supercontinent of Gondwanaland, which began to split up about 160 million years ago. Unusually for such gigantic birds, the ostrich, emu, and rhea all lay large clutches of eggs – up to 30 in the case of the rhea. One reason for this may be that, because the eggs are laid on fairly open ground, they are in great danger from predators. The large clutches not only help secure the future of the species, they provide food for other animals in the ecosystem.

AFRICA The savanna regions of Africa are characteristically dotted with baobabs and thorn trees

AUSTRALIA The open woodland of Australia mainly consists of eucalypt and paperbark trees

SOUTH AMERICA The savanna is known locally as *campo cerrado* or *llanos*, and is home to the rare pampas deer

GRAZING HERDS

Although the savanna's vertebrate grazers all depend on the same foodstuff, grass, each species has slightly different requirements, helping to reduce competition. Animals such as wildebeest and zebra (10C) are migratory, following the rains to take advantage of fresh growth, while Thomson's gazelles are able to survive by grazing smaller territories. Different herbivore species may even unwittingly help one another. Zebra, for example, crop the savanna grass close to the ground, allowing gazelles to reach the succulent, low-growing herbs.

The migration of wildebeest herds is a major feature of the East African savanna. They move around the Serengeti National Park (right) following the movement of the equatorial rainbelt. During the wet season, when the rainbelt is in the south and the growth of grass is at its height, the herds feed on the Serengeti plain. All the females give birth within a few days of each other so that, although the predator populations have a glut of easy victims, they are unable to kill as many young as they would if presented with a steady supply of calves. Near the end of the wet season, when the Serengeti plain is overgrazed, the herds move northwest. They spend the main part of the dry season in the northern Serengeti where the best supply of grass is to be found, before once again making the long trek south.

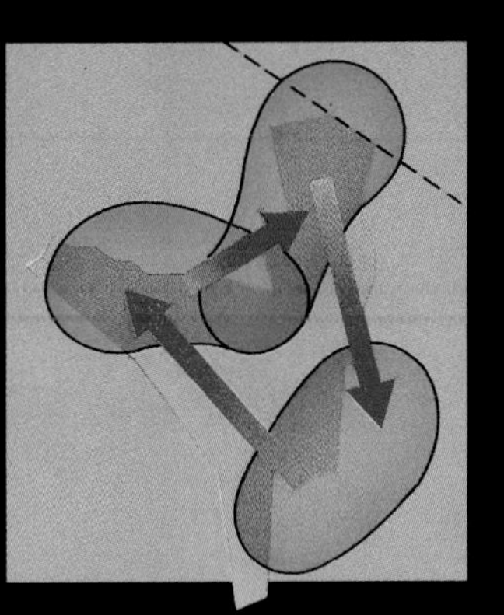

Migration patterns

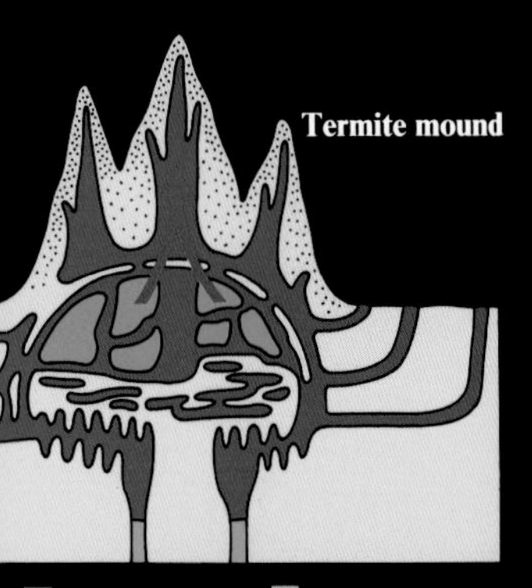

GRASSES

The perennial grasses of the savanna contend not only with seasonal drought but also with frequent burning and heavy grazing. These grasses have well-developed roots and underground rhizomes in which they store carbohydrates, allowing rapid growth after defoliation. In some regions more than two thirds of their total biomass is beneath the surface. The growing points of grasses are numerous and are also situated at or below ground level for protection and rapid regrowth. A savanna tree is much more vulnerable than the surrounding grasses, especially when young, because its growing point is at the top of the shoot, and easily destroyed.

Grasses like *Pennisetum* invest in thick underground stems, but red oat grass employs a different strategy. Each of its seeds has a long corkscrew-shaped awn that expands and contracts in response to the alternate wet and dry periods at the end of the rainy season, screwing the seed into the soil where it remains until the next year's rains.

As the dry season sets in, leaves become drier, more fibrous, and harder for herbivores to digest; many grasses also contain hard silica grains which make them still less attractive to grazers.

Despite these deterrents, plants such as red oat grass are heavily grazed and are usually cropped from their maximum height of 60 inches into dense tufts some 20 inches high.

and scavengers is not so clear-cut. Up to a quarter of all "lion" kills are made by hyenas, with the lions coming along later and scavenging off them.

GRAIN STORE

Most savanna plants flower at the end of the wet season, in July and August. By the time their seed is shed, the ground below has been baked hard, preventing the seed from becoming buried. So at the beginning of the dry season, grain lies loose on the savanna soil (up to 350 lb per acre), and represents a protein- and energy-rich larder for seed-eating beetles, rodents, and birds such as the red-billed quelea, a relative of the communal nest-building weaverbird. Traveling in the opposite direction to the grazing mammals, redbilled queleas form migratory flocks numbering many millions of birds. As the rain belt moves north, the queleas move south, to dry areas where the rains have long since passed and grass seeds are abundant. Many seed-eating birds are also opportunistic predators, and will gorge themselves on locusts and grasshoppers if the opportunity presents itself.

11 | 12 | 13 | 14 | 15

11 | 12 | 13 | 14 | 15

WILD DOG ATTACK

Once the dogs in a hunting pack have sighted a prey herd (usually of wildebeest, zebra, or gazelle) they begin to move slowly forward in a crouching stance. At about 150 feet from the herd, the dogs start to run at top speed (30–35 mph), each one selecting a different victim. When one prey animal falls behind, all the dogs in the pack home in. The leading dog gets a firm hold on the prey and slows it down while the others catch up (7C).

Prey selection

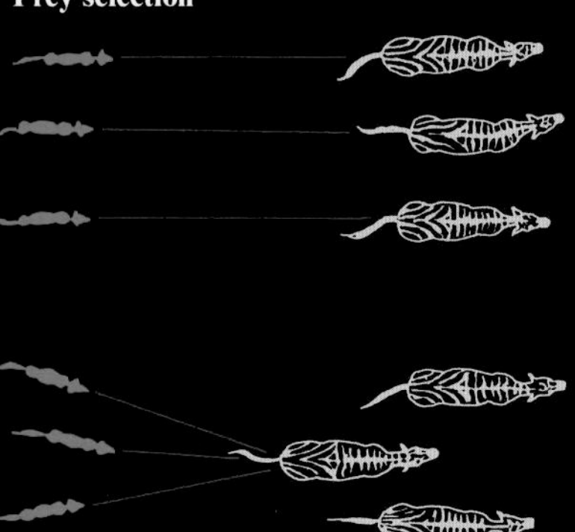

INSECT LIFE

Grass-eating insects are often camouflaged to confound their predators – lizards and birds. Leafhoppers resemble grass seeds, while the migratory locust (15D) changes color with each moult to keep pace with the changing backdrop of drying vegetation. *Macrotermes* termites, which feed mainly on dead vegetation and are important decomposers, build large defensive mounds (14C) out of particles of soil bound together with excrement or saliva. The termites need their food partially softened by fungi, and build large fungus chambers to this end (right), often digging deep tunnels to ensure a suitably moist environment. The fungus thrives by breaking down the dead vegetation and the feces in the wall of the nest, generating great heat as it does so. The porous peaks of the mound act as chimneys, allowing the hot air to rise and carbon dioxide to diffuse to the surface. Fresh air is drawn in through the nest's many tunnels.

SAVANNA DYNAMICS

The key to success

- A wide range of plant life forms and species, supporting a wide range of herbivores in a closely interrelated set of niches. Since most of the herbivores are feeding on one form of vegetation, it is important that each species prefers a different part of the plant
- Intermittent fire
- Long dry seasons, and brief, heavy rainfalls
- The fast and efficient recycling of energy through scavengers

Forces for change

- Failure of precipitation over a period of several years
- A long-term increase in precipitation can also unbalance the savanna
- Overgrazing by fixed and nomadic populations of domestic herbivores
- Overgrazing by wild animals resulting from restriction in available grazing area
- Overutilization of woody species for fuel

Savanna is one of the most delicate of all natural ecosystems. While desert and rain forest are maintained largely by climate, the savanna ecosystem depends on a complex chain of interactions between animals, plants, and the physical environment. If one element is altered, the composition of the community can change rapidly and dramatically until a new equilibrium is established.

The balance of the system may be upset by even a slight change in the climate, for example. Savanna typically receives 10–30 inches of rainfall per year. Yet unless this rainfall occurs with reasonable regularity, the savanna vegetation can easily take on a more desertlike character. The failure of the rains in sub-Saharan Africa in recent decades has resulted in changes and disruption to the savanna vegetation.

Many savanna species are drought-resistant, but even the most resistant cannot survive a succession of dry years, and will eventually begin to wither or fail to seed. In time this forces a change in the species composition of the savanna vegetation, and an overall decrease in tree cover. Serious and prolonged drought may well lead to highly destructive soil erosion in which the crumbling earth is whipped away by the wind or washed away in the heavy rains at the end of the drought.

Savanna interactions

The balance can also be upset by fluctuations in the populations of any of the animal species. Savanna supports a wide range of grazing and browsing animal species and each plays a crucial role in maintaining the balance of vegetation in the ecosystem.

Some species of herbivore, such as the eland, feed on a wide variety of plants. But many feed on very specific plants – or feed in a specific order. As herds of zebra migrate across the plains grazing on tall grass, for instance, they are followed by wildebeest which eat the shorter grass the zebras expose, and the wildebeest are followed in turn by gazelles which graze on the new grass growing in their wake.

This succession plays a crucial role in maintaining the vitality of the vegetation. A survey in Serengeti National Park in East Africa showed how within days of moving in to feed on new short grass after the rainy season, herds of wildebeest had stripped away 85 percent of the biomass. But this heavy cropping stimulated new growth for the gazelles to feed on. Growth in ungrazed areas, however, declined in quality and quantity.

Normally, populations of each savanna species are kept in balance by a whole series of natural checks, but with such a degree of interdependence, it is clear that a change in numbers of any one of them – either up or down – echoes all through the ecosystem. Just how this can work was shown in the 1970s when the dry season in Tanzania's Serengeti National Park became slightly wetter and the dry season slightly drier, with no overall change in the total rainfall.

When more rain falls in the dry season, grass grows throughout the year. First to benefit are grazers such as the wildebeest, since the extra food and water reduce mortality on their arduous dry-season migrations. Scavenging vultures which follow the herd hoping for corpses decline as a result. Competition for food between wildebeest and buffalo is also reduced, and the increase in buffalo numbers favors their main predators, the lions.

Not all herbivores gain from the increase in dry-season grass growth. The extra grass chokes out low-growing herbs that are the main food of Grants gazelles. So cheetahs, which hunt the gazelles, fall in number along with them.

Because the grass remains green and moist all year round, fires become less common, more saplings survive and despite the extra competition from the grasses, many grow into small trees. This, in turn, promotes an increase in the number of browsers such as giraffes.

Human intervention

In recent years, the agents for change in the savanna have been human as well as

Isolated trees and the ecosystem

Although few and far between, trees in the savanna play an important role in the ecosystem by creating special physical conditions around them. In the Arizona grassland, for example, isolated mesquite trees (*Prosopis juliflora*) shade the ground beneath from the blazing savanna sun and cool it significantly. The average yearly temperature of the soil beneath mesquite trees is at least 7°F lower than that of the surrounding open grassland, and in summer the difference can be as much as 20°F. One result is that the soil beneath the tree is considerably moister. It is also much more fertile. This increase in soil fertility is thought to occur because the tree foliage intercepts airborne chemicals. These are then washed in concentrated form into the soil beneath the tree. Although the shade may reduce the amount of light available at ground level for photosynthesis by as much as 50 percent in some cases, the productivity of the grass beneath the tree is nonetheless much greater than that of the surrounding open grassland. Grassland animals benefit from this rich growth, and from the tree's welcome shade. In the African plains, large trees such as baobabs can fill the same roles.

When large numbers of elephants are confined to a small area, food becomes scarce, and they begin to strip trees of bark for sustenance

As hordes of hungry elephants attack trees for food, they can very quickly turn the savanna into a wasteland

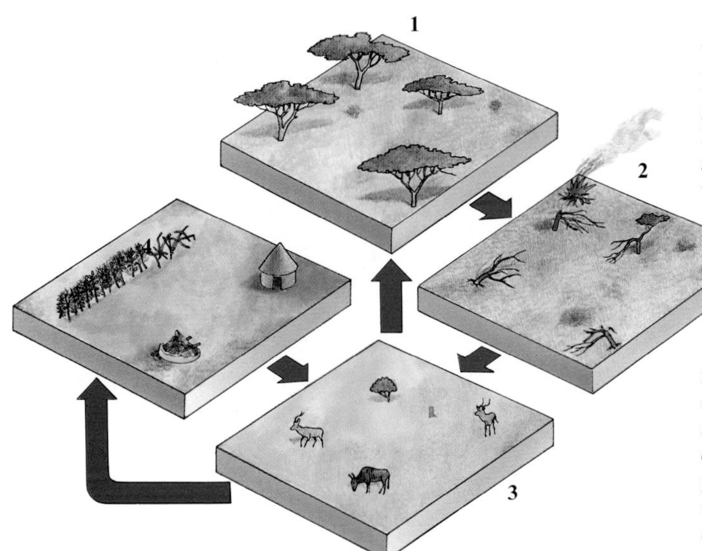

Vandal elephants

Elephants play a vital role in maintaining the balance between bush and grass in East African savanna.By browsing on trees and shrubs, they keep the bush in check, leaving plenty of open grass for grazers such as zebras. But in Kenya's Tsavo National Park in the 1960s, elephants were confined to areas where they were protected from poaching and hunting, and their numbers increased rapidly. Food became scarce as the elephants multiplied. Some began to dig down for the roots of the trees, causing soil disturbance. Then they stripped the bark off acacias and baobabs (1) and pushed trees over to get at upper foliage and young shoots – initially attacking only a few types of tree, but soon virtually all. As the land dried out, fires destroyed more trees, leaving an open landscape with just a few stumps (2). The elimination of trees benefitted grazing wild animals such as zebra and wildebeest but also encouraged the introduction of domestic cattle (3). Larger species which feed predominantly on tree foliage, such as giraffe, rhinoceros, and of course elephants, have either to migrate out of the Park area – in which case they may cause severe damage to farm crops (4), and risk being shot for their pains – or die of starvation.

natural. For thousands of years, the traditional nomadic herding communities have fitted well into the savanna ecosystem, continuously moving their cattle to new pastures and in this way avoiding the dangers of overgrazing. Nomadic herding requires large, open spaces. However, economic demands and rising population have ensured that more and more domestic cattle are kept permanently in one place – in vast numbers. When kept in one place, cattle have a devastating effect on the savanna ecosystem. In feeding, they select the species that they prefer and leave aside tougher plants. Their selection of fine grasses lets weedy invaders move in, and the nutritional value of the forage gradually declines. The hungry animals then strip the ground bare, pulverizing the soil with their hooves. Wind and rain carry away the loose topsoil, irreversibly reducing the productivity of the land, on which only tough, drought-adapted shrubs can survive.

It is not only grass that is disappearing – the farmers frequently cut down trees for fuel and hew up the less woody plants as further food for their animals. The ecosystem is thereby rendered even less stable.

Within the savanna lands of East Africa there are many parks maintained by national governments. Their natural beauty and abundance of wildlife attract millions of visitors a year. But even these sanctuaries need careful management if the diversity of wildlife they contain is to be maintained. Tanzania's Mkomazi Game Reserve, for example, has lost four out of 43 large mammal species since its establishment in 1951. With an area of barely 400 sq miles, the park is too small to support animals that require a large territory – especially predators. Cultivation of neighbouring land also drives many animals into the reserve, disturbing the balance of the community. At over 400,000 sq miles, Kenya's Tsavo and Tanzania's Serengeti provide a more stable habitat for large animals, but even here populations of rhinoceros, cheetah, and elephants are under serious threat from poachers.

BAT POLLINATION

Although some plants, like *Nothofagus*, are wind-pollinated, the dense foliage with its resulting still air means that a high proportion of New Guinea's plants are pollinated by birds, insects, and even bats. Compared to the bird pollinated flowers, which are brightly colored but odorless, a bat-pollinated flower often smells powerfully, and is pale to make it more visible in the dark. The flowers of *Parkia* hang pendulously above the canopy, making them easy for the bat, with its upside-down roosting technique, to reach while simultaneously discouraging other animals. New Guinea has more than 50 species of fruit bat (2A), some with wing spans of over 3 feet, and many of them are pollinators. It is only the smaller insectivorous bats that echolocate: fruit bats rely on their large eyes and superb vision to navigate.

Lurking in the branches of the ascendants are birds of prey such as the bat-eating hawk (1B), waiting to swoop down and pluck some unsuspecting creature from the rich larder of the canopy.

6 | 7 | 8 | 10

6 | 7 | 8 | 9 | 10

BRANCH TO BRANCH

Negotiating the canopy in search of food – both vertically and from tree to tree – is an important skill in the rain forest. The sugar glider (3B) has flaps of skin stretched between its fore and hind limbs which it spreads out and uses like a parasail to glide up to 50 yards from tree to tree in search of sap and gum. As the sugar glider falls, its motion creates a powerful air flow in the opposite direction. By angling its body and curving the surface of the skin-flap, the sugar glider maximizes the resulting lift. By curving each flap a different amount it can even steer, and it ensures a soft landing by tilting at a more severe angle at the end of its flight, generating turbulence, which drags it backward, and creating a controlled stall.

The green python is a master climber, as is the cuscus (11E), an animal which spends all its time in the trees eating leaves, fruit, and flowers. Like the New World monkeys it has a prehensile tail, which in the cuscus is naked for its last two-thirds, with rough bumps on the inner surface to provide grip.

New Guinea's is a unique rain forest in that it has no primates or prosimians (such as lemurs and bushbabies). Apart from the bats, all the tree-dwelling mammals of New Guinea are marsupial. Even the kangaroos took to the trees after leaves and fruit. Tree kangaroos (12B) evolved smaller hind legs and the ability to move them independently of each other, which their Australian cousins cannot do. They also developed stronger arms and a better grip, but they remain slow and cumbersome climbers, and really only survive because of the lack of competition.

ANT-PLANTS

An epiphyte, such as *Myrmecodia* (10E), grows on trees but only uses them for support. It is not a parasite and takes no nourishment from them. Nevertheless a host tree that is cleared of epiphytes will often show a marked increase in growth-rate. The *Myrmecodia* is also known as the ant-plant, because it has a tuber which is filled with interconnected chambers, which provide a home for ants. In return, the aggressive insects help to deter other animals from feeding off of the plant. They also provide a large part of the plant's nutrition. Some of the internal chambers have absorptive warty surfaces, on which the ants deposit droppings, dead ants, scraps of food, and fungal spores.

Ecosystem Profile *The New Guinea Rain Forest*

The 300,000 sq miles of New Guinea, balanced at a point where the Asian, Australian, and Pacific tectonic plates have collided, contain some of the grandest mountains and most violent volcanoes in the world. Over two-thirds of the land area is covered in dense tropical forests, which contain some wholly unique animals. Primates, cats, pigs, and deer have never existed here, and a bizarre range of marsupials and birds have developed to fill the ecological niches that these creatures would occupy in any other rain forest in the world.

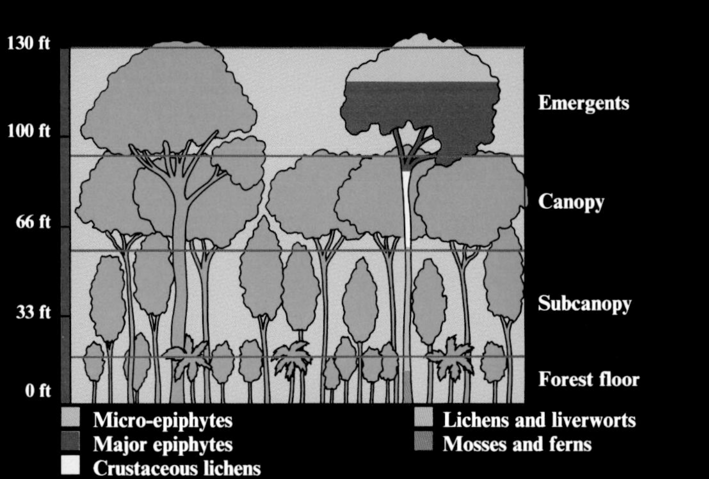

FOREST PATTERNS

The vegetation of the rain forest is so dense, and so competitive, that it organizes itself into distinct layers as the many plants occupy their particular niches in the struggle to survive. The tallest trees, such as *Araucaria (*1A*)*, emerge through the thick canopy formed by trees like rosewood and taun. Because the canopy gets so much light, it is the site of the most luxuriant fruits and flowers, and the animals they attract. Little sunlight penetrates to the subcanopy trees, such as the nutmegs, and even less to the shrubs and ground-level vegetation. This leads to a wide vertical range of microclimates: the air near the ground is perpetually saturated with moisture, while in the upper canopy evaporation can be as high as in the savanna, and many plants and leaves growing here are adapted to surviving practically drought conditions. The air plants, or epiphytes, growing on a single emergent tree, reflect this span of different habitats. The exposed crown is inhabited by microepiphytes. The upper limbs house most of the large epiphytes, which are adapted to semiarid conditions with much sunlight. Many species of epiphytic orchids (15E) and some ferns, such as *Pyrrosia* (10E), are like desert plants in that their stomata, or ventilation pores, remain closed during the day and only open at night, when dehydration is less. The dry upper part of the bole has crustaceous lichens, while the moister lower portion has lichens, liverworts, and mosses. The base of the tree is covered with mosses and filmy ferns. Winds often rub branches against each other, and branches of a softer wood are literally sawn off by a harder. As a result, arid-zone epiphytes often crash into the moister regions below, where they must adapt or die. This unique, sudden ecological stress may be one of the evolutionary factors behind the rich variety of rain forest plants.

Although the goal of climbing epiphytes is to scale the trunk of a tree into the well-lit canopy, many climbing shoots are at first negatively phototropic – they actually seek out the deepest shade, because this is likely to be caused by a tall, suitable host.

RAIN FOREST

Under the shade of the towering, liana-festooned trees of the tropical rain forest is the richest, most astonishingly diverse ecosystem on Earth. A perpetually warm, damp climate ensures that plants grow bigger and more luxuriantly here than in any other environment. All kinds of animals and plants thrive in this fruitful place. Over 1.5 million species of plant and animals live in the tropical rain forests, more than in all the rest of the world – and many more have yet to be identified. Over 300 different species of tree alone can be found in just 1 sq mile of the Brazilian forest.

Yet despite this fecundity, the rain forest environment is remarkably fragile. Once cleared for agriculture, the land takes many hundreds of years to recover. Vast areas of the tropics in the Amazon and Congo basins and in the East Indies are still covered by virgin rain forest, but an area the size of Ireland is destroyed every few years.

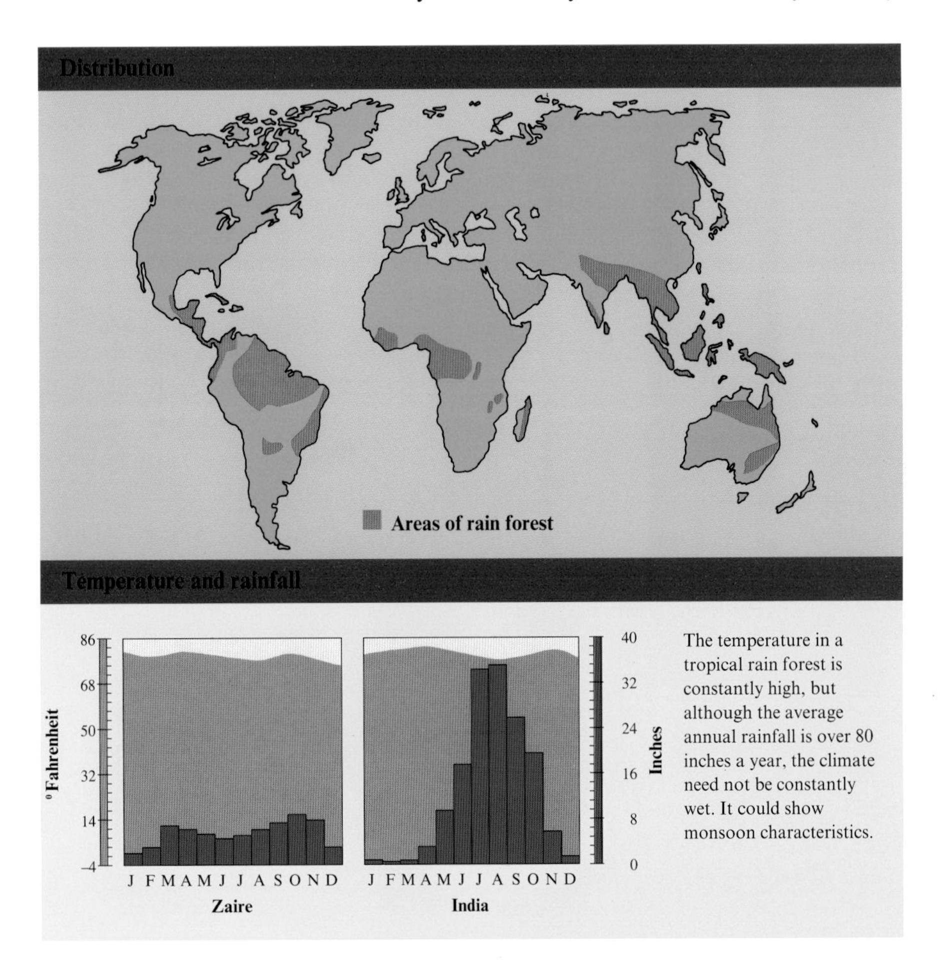

The temperature in a tropical rain forest is constantly high, but although the average annual rainfall is over 80 inches a year, the climate need not be constantly wet. It could show monsoon characteristics.

A rain forest may extend from lowland mangrove swamp, through montane forest right up to the high-altitude, alpine vegetation

The impenetrable jungles which are such a feature of films shot in the tropics bear little resemblance to the mature rain forests of the lowland tropics. These have an open understory because little light penetrates to the forest floor, most of it being trapped by the trees of the canopy. It is from above that the tropical forest appears to be an almost completely uniform blanket of green. Here and there, however, the umbrella-shaped crowns of a few very tall trees emerge clear above this undulating sea of leaves. These forest giants, called emergents, soar some 130–160 feet above the forest floor, yet they are rarely more than a few hundred years old, for trees reach maturity and die rapidly in this incredibly rich environment. The moist warmth of the rain forest means a tree can grow 80 feet in height and more than 16 inches in girth in just five years. Old trees are continually crashing to the ground after rotting or being struck by lightning. The gap they create as they plunge through the undergrowth does not last long: it is rapidly filled by species actively competing for the light.

Beneath the crowns of the emergents is the dense, almost continuous main canopy, formed by the mop-shaped tops of trees 80–120 feet tall. Some 70–80 percent of the Sun's light energy is absorbed by the main canopy, which also helps trap moisture and shelters the forest from the wind. Underneath it is dark, humid, and still. Despite the shade, there may be a thick understory of slender trees and saplings growing at varying heights. But right near the ground there is very little growth – although on riverbanks, on hillsides, and in clearings where light penetrates more easily, there may be a thick undergrowth of grasses, ferns, herbs, and saplings at ground level. Even on the dark forest floor, though, there are saprophytic flowers that obtain their energy not by absorbing light but from the abundant leaf litter on the forest floor.

There is barely a tree in the forest that is not festooned with the stems of lianas and creepers, which wind up through the branches toward the light. Nor are there many trees free of epiphytes and parasitic plants that root directly in the tree, not in the ground. The roots of epiphytes such as bromeliads are not for absorbing water, but for clinging to their lofty perches; they take in water through special scales whenever it rains and survive in between showers by storing water in their succulent leaves.

Each of these different parts of the forest has its own fauna, though many insects, reptiles, birds, and mammals move freely up and down through the forest. High above in the sunny crowns of the towering emergents, though, live many creatures that rarely, if ever, venture down into the shadows below, including agile insect-eating birds such as the great-eared nightjar and fruit-eaters such as the hornbill and the spectacularly beaked toucan. Here, too, live some of the world's largest birds of prey, such as the South American harpy eagle and the Philippine monkey-eating eagle, which swoop down on monkeys in the canopy below. The canopy itself teems with animals, all adapted to life far above the ground. One survey identified 600 different species of beetle alone in the canopy of a single tree. The majority of the different species in a rain forest are to be found in the canopy.

In the shady, less populous branches of the understory lurk snakes and big cats, such as the jaguar and ocelot in South America and the clouded leopard of Southeast Asia, ready to drop on to their unsuspecting prey, which may be anything from small mammals or reptiles to tapirs and peccaries.

Down on the forest floor, elephants, tapirs, deer, gorillas, and other large mammals forage on the leaves of the understory, along with flightless birds such as the bowerbird.

One of the remarkable things about the rain forest is the extraordinary degree of interdependence between all the different species. Flowering plants, for example, are rarely pollinated by wind but rely on insects, bats, birds, and even mammals to spread their pollen.

Perhaps the most delicate aspect of the rain forest is the recycling of nutrients. The forest may be luxuriant but its soils are actually very poor in nutrients, even though they are very old and deep. The luxuriance depends on the way huge quantities of leaf litter fall to the forest floor all the time – typically more than 4 tons on every acre every year. Woody tissues are gradually broken down by termites, and softer organic matter is broken down by mites and insects and the fungal mycorrhiza, releasing the nutrients to be taken up through small spreading plant roots. If the supply of litter is interrupted – perhaps by the destruction of the trees – the soil loses its fertility for ever.

A rain forest profile

The nature of the rain forest varies with altitude, and above 1,600–2,300 feet the flora and fauna undergo a marked change. In New Guinea this zone houses the oaks and Lawson cypress, kauri pines, and towering hoop pine and klinki trees (both related to the monkey-puzzle tree) of the great highland forests. Above these are the mid-montane mist forests, where 100-feet *Nothofagus* (southern beech) trees tower over an enormous variety of ferns and orchids.The upper montane forest is dominated by short (50–65 feet) podocarp conifers of southern origin. There is less species diversity than in the mid-montane zone. Every surface is covered with wet mosses, lichens, ferns, orchids and other epiphytes, and the forests are often shrouded in mist all day. At 10,000 feet, the forests are dominated by only two or three podocarp conifers or by *Papuacedrus*. Members of the heather family are common and tongues of sub-alpine grassland penetrate the forest from higher altitudes.

A permanent mist shrouds the stunted trees of elfin cloud forest

Ecological counterparts

Rain forests throughout the world have niches that are occupied by agile, climbing mammals. On the mainland continents of South America, Africa and Asia, the most successful of these are the simians: the apes and monkeys. On large, isolated islands, such as Madagascar, or some of the islands of Indonesia, which monkeys have never had a chance to colonize, the more primitive prosimians, the lemurs, lorises, and tarsiers, may be the most highly evolved non-human life. What many of the most agile of these creatures share is a long tail, which may simply be used for balance, as in the lemur and the Old World monkeys, or may be prehensile, and capable of gripping, as in the monkeys of the New World and the opossum, of Australia and the Americas.

Spider monkey

Lemur

Opossum

decorated with stones, shells, fruits, dead insects, and other objects – to attract a mate. The bower birds and the most striking birds of paradise are polygamous: males try to attract and impregnate as many females as possible.

Many of New Guinea's insects are gargantuan, like the stick insects (6D) that can be more than a foot long. The largest known insects are the bird wing butterflies (15A), with some species having wing spans of over 1 foot. The butterfly lays its eggs on *Aristolochia* (Dutchman's pipe). The caterpillar eats the leaves and in some species ringbarks the stem before pupating, to kill off any leftover leaves and reduce competition. Disturbance does not threaten the butterfly as *Aristolochia* regenerates opportunistically in forest clearings. Indeed, around Mount Lamington the high numbers of butterflies may be the direct result of high volcanic activity.

LEAF STRUCTURE
Leaves function differently at different heights in the forest. In the dry upper levels they may form a cup to retain water. In the moister lower levels the leaves are designed to get rid of moisture quickly, before fungi can attack, and to allow transpiration – where water evaporating from the leaf surface sets up a chain reaction that draws mineral-rich water up from the roots. Up to 90 percent of the leaves in the forest undergrowth have drip tips – narrow downward-pointing funnels to drain off water (14A). It has been found that a leaf with its drip tip cut off can take five times longer to dry. Millions of these leaves together act like a giant showerhead, diffusing a heavy rainfall by reducing the size of the droplets that reach the ground and spreading them evenly over the surface. In this way they reduce soil erosion in the forest.

Light is scarce underneath the canopy, and plants must make the most of it. The undersides of begonia leaves (6E) have red pigments to make use of red light that has been backscattered from surrounding vegetation. Many leaves can turn to follow any available rays of light.

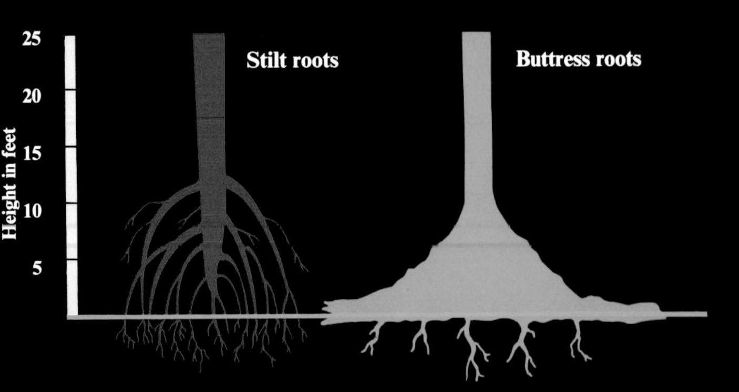

SEEDS AND ROOTS
Although the seeds of some canopy trees, like the rosewood, are wind-dispersed, most seeds in New Guinea are dispersed by animals. Fruit-eating pigeons spread the *Lauraceae*, bats spread the *Terminalia* and cassowaries spread the *Pometia pinnata* or taun. The seeds of the taun have a fleshy outer coat, and if they fall straight to the ground they may have only a 5 percent chance of germinating. If they pass through a cassowary's gut, which strips away the fleshy coat and excretes the seed some distance from the shade of the parent, in a ready-made puddle of fertilizer, the chances of germination rise to 95 percent.

A tree must then cope with the problems of growing in shallow, unstable soil. Many trees, including *Myristica*, grow stilt roots, or air roots. These help spread the weight of the tree over a wider area. In the often flooded and poorly aerated rain forest soils, they are also better at taking up nutrients than the long underground root systems which often suffer oxygen depletion. Other trees, such as *Tetramales*, develop heavy buttresses above the ground to support them.

The leafless woody trunks of trees in the rain forest often have flowers or fruits growing directly out of them, a condition known as cauliflory. This makes them conspicuous to pollinators and seed distributors.

18 19 20

F
G
H

JEWELS OF THE FOREST

A group of male raggiana birds of paradise (9B) will have a favorite display tree high in the canopy. They will peck away many of the leaves that might obscure them, and here they will perform their competitive communal display, or lek. This extravagant show involves frantic wing-flapping and tossing around their bright golden display feathers, as well as hanging upside down in a shimmering frenzy. Any nearby female will choose her mate on the quality of the performance. There are nearly 40 other species of birds of paradise on New Guinea, though few quite as spectacular as the raggiana. By choosing the showiest male the females increase their chances of having showy male offspring which will in turn be a likely choice for other females. In this way they improve their chances of passing on their own genes. The same is true of the related bower birds (18G), which are mostly dull to look at but build complex bowers –

11 | 12 | 15

11 | 12 | 13 | 14 | 15

Other epiphytes have different methods of absorbing nutrients, for example from the humus of decaying leaves, twigs, bark, fruits, droppings, and animal remains that accumulates along tree branches. Parasites, in contrast, feed on their hosts. The *Rafflesia* (19H), lives entirely within the root structure of *Cissus* vines, exposing only its flower. *Rafflesia*'s football-sized bud can take years to develop, and then the 3-feet-wide flower – the largest in the world – is only open for a few days. It looks and smells like a rotting carcass, in order to attract the flies that pollinate it.

Ant-plant chambers

Entrances and passageways

THE FOREST FLOOR

The largest flesh-eating animals in New Guinea are reptiles like the 16-feet-long, tree-climbing Salvadori's monitor. The big cats and other mammalian carnivores that stalk the floors of most rain forests never evolved here, so the ground teems with flightless birds like the cassowary (20E), and flying birds like the fruit pigeons and bower birds that also choose to feed or nest here. This is not to say that the cassowary would be defenseless even against large predators. It can grow 6 feet tall and its feet are equipped with long spurlike claws. The cassowary is an adept kick boxer, and can easily disembowel a man.

The largest carnivorous marsupial is the domestic-cat-sized quoll (16H). It mostly feeds on birds, insects, small mammals, and reptiles, though it will attack prey larger than itself. Yet the biggest flesh-eating mammal on this island of marsupials is the insectivorous echidna (16E). Also known as the spiny anteater, the echidna is a monotreme, like the duck-billed platypus, and shares with the platypus the distinction of being the only mammal to lay eggs. The echidna has long spines all over its body, and when threatened will curl up like a hedgehog. Its long thin snout is ideally designed for poking into the tunnels of termite mounds or ants' nests, but means that if it catches a worm, another favorite prey animal, it must begin eating at one of the ends. The echidna does not suck up the worm, but harpoons it with a long, barbed tongue and reels it in.

RAIN FOREST DYNAMICS

The key to survival

- The rain forest's very instability is what preserves it – through the rapid turnaround of species and resources. Tree falls, creating clearings, and a constant pattern of other natural disturbances are essential for the maintenance of rain forest diversity.

Forces for change

- Fires, either natural or manmade
- Volcanic eruptions and the tons of ash they throw into the air
- Earthquakes
- Cold air draining down from mountain tops
- Tree falls, sometimes caused by lightning
- Droughts
- Flooding
- River meanders, shifting back and forth and submerging vast areas of rain forest
- Logging and road building, and the use of timber for fires and cooking
- Clearing for farmland

Rain forest ecosystems are in a constant state of turmoil. Plants grow with astonishing speed in this hothouse environment, and trees grow faster than anywhere else in the world. This rapid growth, though, is matched by comparatively short lives. Individual rain forest trees rarely live more than 300 to 400 years, compared to those of temperate forests which can live for more than 1,500 years.

There is a purpose behind all this frantic activity. The survival of the rain forest depends on its ability to conserve nutrients, and to cycle them very quickly within the ecosystem. In temperate forest, nutrients are retained in the soil until they are needed. But the soil beneath the rain forest is often poor and deeply leached, and trees and other plants depend on retrieving nutrients from plant litter before they are washed away forever.

Many nutrients washed from the surface of old leaves fall in droplets to the forest floor where they are rapidly reabsorbed by the fine root hairs which form a mat at or just below the soil surface. If nutrients are permanently lost by large-scale destruction, either naturally or as a result of extensive logging and subsequent fires, or large-scale clearance for agriculture, it can take up to 700 years to regenerate a balanced ecosystem.

Filling the gap

When a gap is created in rain forest, it is essential that nutrients are not washed away by rain. Tree falls are a constant feature of many rain forest communities and the ecology of the rain forest plants is adapted to rapidly filling any such gaps and thus reabsorbing nutrients that have been made available by the fall.

There are always plenty of seedlings ready to fill in a gap. Some species have long-lived seeds which germinate only when exposed to bright light; others have very light, wind-dispersed seeds which can rapidly colonize any open space. Within a year of a gap appearing, seedlings of species such as the swamp *Terminalia* can grow up to 23 feet. In five years or so, the rain forest eucalypt, *Eucalyptus deglupta*, can grow to a height of 80 feet with a trunk diameter in excess of 12 inches.

Small gaps are created in the rain forest quite naturally all the time. Lightning strikes, for instance, are frequent in the tropics. In ordinary rain forest, they tend to kill single trees. But in mangrove forests, they can kill several at a time. This is because the lightning is conducted along the mangrove roots to other trees growing from the same root system, killing trees far away from the original strike. The result is that many mangrove forests never reach maturity.

The tinder box

Every decade or so a warm current called El Niño flows in the Pacific, affecting climate right through the tropics. The most frequent causes of large scale destruction around the Pacific are the droughts associated with El Niño. Droughts alone are capable of causing the death of plants, but their major consequence is the frequency of rain forest fires. The extent of the damage can be gauged by the fact that during the 1880s the dense smoke resulting from extensive forest fires caused a complete halt to coastal shipping along the seas north of New Guinea, when visibility was cut to a few yards. Charcoal is common in many rain forest soils throughout the tropics and is a constant reminder of the effects of drought and fire. Many tree species go into "crisis flowering" following these events and produce a copious crop of seedlings which provide the source for the new generation of rain forest individuals. In 1982 the most destructive fires on record occurred in northeastern Borneo, with critical consequences. Before the fire there had been an extensive period of construction of large plymills based on the huge, largely unexploited dipterocarp forests of East Kalimantan. It has been estimated that over 80 billion US dollars' worth of forest was destroyed in the fires, which covered thousands of acres.

Bromeliads and tree frogs

Bromeliads are epiphytic plants that live in the rain forest canopy. To see themselves through periods of drought these plants have evolved closely fitting, overlapping rosettes of leaves, which effectively trap water into standing pools. Many animals, such as birds and monkeys, drink from the bromeliad, and almost never need to venture to the ground. But the bromeliads are much more than mere drinking vessels. Lizards, snakes, and even crabs and frogs are able to make the ponds in the bromeliads' leaves their home.

Tree frogs can spend their entire life-cycles here, feeding on insects and larvae in the water, and even laying their eggs inside these watery nests yards above the ground, and miles from the nearest river. The female marsupial frog actually carries its eggs and developing young around in a pouch on its back until it finds a convenient bromeliad, into which it tips the tadpoles.

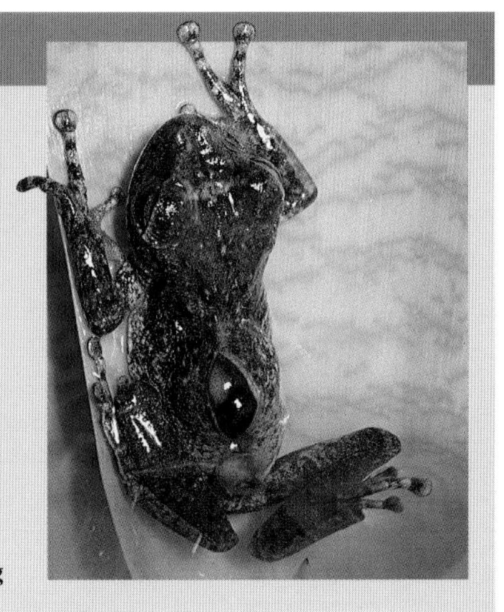

Marsupial frog releasing young

Earthquake damage in a rain forest valley

Cold air drainage

At high altitudes in the rain forest, extensive grassed valleys, often dotted with tree ferns, are a common sight. In their natural state these valleys would be clothed in montane forest consisting of large trees and thick undergrowth (1). Destruction of the vegetation by earthquakes (which often occur in many areas of rain forest), droughts, and fires creates extensive gaps in the canopy (2). Cold air from surrounding mountain peaks drains down the valley system (3), killing seedlings from the surrounding forests and creating a micro-climate which mirrors the conditions at higher altitudes. Continuing disturbances result in a gradual expansion of the gaps and increasing occurrence of grassland and forest margin species. When fully developed, these regions include examples of all the plant and insect species of the subalpine grasslands (4). Genera such as *Ranunculus*, *Anaphalis*, *Coprosma*, and the tree ferns of the genus *Cyathea* are all inhabitants of these distinct areas of vegetation formed by cold air drainage. A similar effect is commonly seen wherever cold air drains down along a stream, with subalpine species growing along the stream banks.

1

2

3

4

Regenerated rain forest, dominated by tree ferns and other species typical of high altitudes

Rhododendron flowers

The clear skies which occur during El Niño cause damage and destruction at higher altitudes by increasing the likelihood of freezing temperatures. Seedlings of tree species along the forest margins and within mid-altitude grasslands are killed by the resulting frosts, as are the poorly adapted forest margin species such as the mid-altitude tree ferns and shrubs.

Hurricanes are also more common during El Niño. In particular, western Pacific Islands such as Eastern Papua New Guinea and Fiji, as well as the Caribbean Islands and adjacent areas of tropical America suffer from frequent hurricanes. In mountain areas of the tropics local cyclonic winds often result in extensive destruction of ridge-top forests, and many montane *Castanopsis* forests owe their origin to local wind storms and the interference of man.

Heavy rainfall can also be destructive, particularly in forests growing on mudstone. The weight of the forest with its underlying soil saturated by rain can exceed the soil strength and whole areas simply slip away in massive landslides. But rainfall is not the only cause of flooding, and the meanderings of larger rivers are a continuous source of rain forest destruction.

Tropical rain forest often occurs in geologically unstable areas, prone to earthquakes and volcanic eruptions. The effects of volcanoes are felt not only in the blast areas where complete destruction of the vegetation can occur, but also in the often extensive areas where the forests are showered by volcanic ash. Areas with constant volcanic ash falls traditionally have rich soils, though the floras may nevertheless be limited in extent because they are covered with ash which offsets the effects of the fertile earth. In Borneo the areas of volcanic ash-rich soil support a completely distinct group of species.

between 11,500 feet and 14,800 feet high, while the guanaco both grazes and browses, which allows it to survive in a greater range of habitats, including lowland forest. Although the vicuna needs to drink, the guanaco gets all its moisture from its food. The guanaco is also much larger than the vicuna, and favors lower altitudes, though the ranges of the two overlap between 11,500 feet and 13,000 feet. Neither species is naturally migratory.

It is possible that the llama and alpaca are descended from wild guanacos, or that the alpaca is the result of crossbreeding between the vicuna and the llama. Clearly all four species are closely related. Any possible combination of pure or hybrid crosses will produce fertile young. All the cameloids breed seasonally, to coincide with the summer explosion of plants.

Cameloids have been domesticated in the Andes for more than 4,000 years. Their wool, meat, milk, burnable feces and, in the case of the llama, surefooted pack-carrying across difficult ground have been essential to the survival of man in this testing environment. Llamas were the economic heart of the Inca empire. It was only when the Spanish arrived, bringing guns and sheep, that cameloid numbers began to decline, and even today one of the main threats to the guanaco is competition from domestic animals.

KEEPING WARM

One of the most important adaptations by animals in high mountains is an ability to reduce heat loss. Large mountain animals, such as mountain sheep and vicuna (5G), have fur that is very efficient for this purpose. Small rodents like the chinchilla (9E) and viscacha (10F) are much less well insulated, and have to burrow in the soil or under the snow to minimize heat loss.

The guanaco has thick fur on its back but a much thinner covering on its underside. It exposes this thinly insulated area to the sun to warm itself up when appropriate, and when the sun is not shining and temperatures are low it curls up so that only the thick fur shows on the outside.

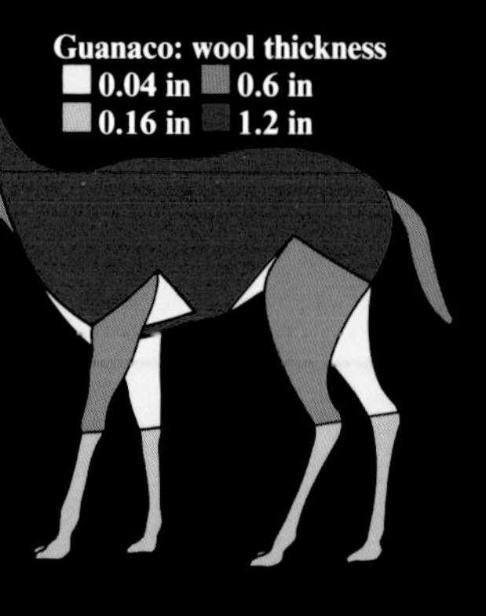

The vicuna and alpaca are particularly prized for their thick, soft wool, as a result of which the vicuna's numbers have been under threat from hunting, while the domestic alpaca has thrived.

6 | 7 | 8 | 9 | 10

6 | 7 | 8 | 9 | 10

WATER BIRDS

With their characteristic bright pink plumage, flamingoes are distinctive and familiar inhabitants of Africa, the Caribbean, and Europe, where they have been known for many years. However, it was not until 1957 that the first colonies of three Andean species were discovered in Bolivia. They live 13,000 feet above sea level on lakes which are saltier than the sea. They feed by filtering small crustacea from the water that they sluice back and forth through their sievelike beaks. The Chilean flamingo is similar in size to the European and African species but has longer wings, thought to be an adaptation to assist flight in the rarefied air. The other two species, the Andean flamingo (2F) and James's flamingo, are both smaller and prefer to remain at the high lakes throughout the year, whereas the Chilean flamingo moves to lower altitudes to escape the bitter cold of the winter. Although the lakes are fed by warm volcanic springs, it is so cold at such high altitudes that even these saltwater expanses

The Andes stretch the whole length of South America, from Lake Maracaibo in the north to Tierra del Fuego in the south. The many peaks, including Mount Aconcangua, which is nearly 23,000 feet high, make up the longest mountain chain in the world. The Andes are an effective barrier between the coast and the South American interior, dividing species of plants and animals. As the mountains rise they display a staggering range of vegetation, from lowland rain forest to grass plains to barren, icy wastes. Yet the inhospitable slopes contain many surprises: flamingoes, lizards, and even toads thrive in the cold, rarefied air.

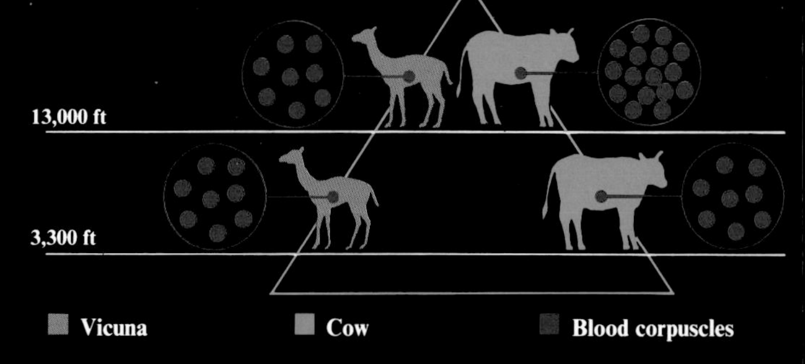

CAMELS

Outside the tundra, the members of the camel family are among the main large herbivores of the arid regions. The dromedary is found throughout North Africa and Southwestern Asia, and has been introduced successfully to Australia. The two-humped bactrian camel still lives wild on the Asian steppes. And four species of cameloids live in South America: the presence of the wild guanacos (4F) and vicunas (5G), and the domestic alpacas and llamas, reveals just how many characteristics the Andean mountains share with other, lower, arid regions.

All of the world's camels have certain characteristics in common including unique elliptical red blood corpuscles and an isolated upper incisor tooth, although only in the cameloid males does this tooth become sharp and tusklike for use in fighting for territory. Dominant male vicunas and guanacos occupy a territory with several females and young offspring.

Cameloids differ from dromedaries and bactrian camels in that the pads of their toes are not as wide, and are independently movable, which makes them more surefooted on rocky slopes. Like their Old World relatives, the cameloids are typically resilient in conditions of drought. They produce very little, highly concentrated urine and extremely dry feces; they eat thorns and dry vegetation that other mammals reject, and are aided in digesting this fibrous material by their habit of chewing the cud.

RESPIRATION AT ALTITUDE

Animals taken up to high altitudes at first suffer severely from oxygen deficiency but gradually acclimatize by increasing their rate of respiration, their heart rate and the number of oxygen-transporting red corpuscles in their blood.

Curiously, the animals best adapted to high altitudes – such as the cameloids and the yak – do not show any increase in red corpuscles, although they do have comparatively larger hearts and lungs to start with, and they do increase their respiration and heart rates at high altitude.

The vicuna is a grazer, feeding mainly on the *puna* grassland

1 2 5

D

E

3 4 5

MOUNTAINS

Mountains are surely among the most spectacular environments in the world. Precipitous cliffs and jagged ridges may soar up to snow-capped peaks beyond the reach of any creature but the most adventurous birds. Yet what makes mountains remarkable from an ecological point of view is that each has its own unique climate. A mountain may have its foot in tropical seas and its head in arctic winds. Or it may be sunny and warm on one side, and wet and bitterly cold on the other. On some mountains, conditions can change from equatorial to polar within a few miles.

Each mountain, then, no matter where it is in the world, has its own particular ecology and its own special collection of plants and animals – often bearing scant resemblance to the ecology of the region in which it is located. High on mountains in Europe and North America where temperate forest is the norm, a tundra ecology may flourish, while mountains rising above barren tropical deserts may be home to lush rain forests or even alpine flowers.

Climbing a mountain is in some ways like traveling from the equator to the poles. Just as tropical flora and fauna give way gradually to temperate, boreal, and arctic flora and fauna moving away from the equator, so ascending a mountain takes one through various vegetation zones. In the tropics, vegetation may change from rain forest in the foothills to tundra at the summit.

However, every great mountain creates its own set of conditions, and the changes with altitude on a mountain do not exactly reflect changes with latitude. An ascent of Mount Kilimanjaro in East Africa, for example, takes the climber from wide open savanna up through damp, dense forest, stands of bamboo, and coppices of shrubs and dwarf trees to alpine grassland and, eventually, bare rock at the summit.

In the eastern Himalayas, on the other hand, the climber starts amid subtropical forest and moves up through woods of deciduous trees mixed with bamboo and grass. Above 5,000 feet, the Himalayan woods are almost like cool temperate forests, with oak, chestnut, maple, alder, birch, laurel, and magnolias. By 9,000 feet, the deciduous trees have given way to conifers and junipers, interspersed with willow, bamboo, and rhododendrons. Above 10,000 feet, the woods thin out, and by 12,000 feet, the climber emerges on to open alpine meadows, with small herbs and tiny, colorful tundra plants and lichens giving way, eventually, to bare rock and snow.

All these different levels of vegetation reflect the fact that the air becomes colder naturally with height. The rate at which the air cools varies, but the temperature drops by 0.7°F every 200 feet. Above a certain level, known as the snow line, conditions are so cold that the mountain is almost permanently covered by snow and ice – except in places too steep for snow to cling to. In the tropics and near the sea, the snow line can be as high as 16,500 feet, but drops to 9,000 feet in the European Alps and right down to sea level at the poles.

Mountains are not only colder than lowlands but often wetter too. For example, the damp conditions encourage very luxuriant growth on the lower slopes of tropical mountains. However, because sunlight can more easily penetrate the leaf canopy to the undergrowth, the trees are often shorter and covered in a profusion of mosses and lichens and abundant ferns and climbers. Higher up, tropical mountains may be swathed almost permanently in cloud, creating a strangely tangled kind of tropical rain forest known as cloud forest, full of mossy, stunted trees with contorted branches. On the Ruwenzori in East Africa, dwarf trees mingle with giant ferns, groundsels, and lobelias as big as trees.

Conditions on a mountain, and so the nature of the biome, vary not only with height but also from one slope to the next. There is often a dramatic difference in the range of species between the sunny side of the mountain and the side always

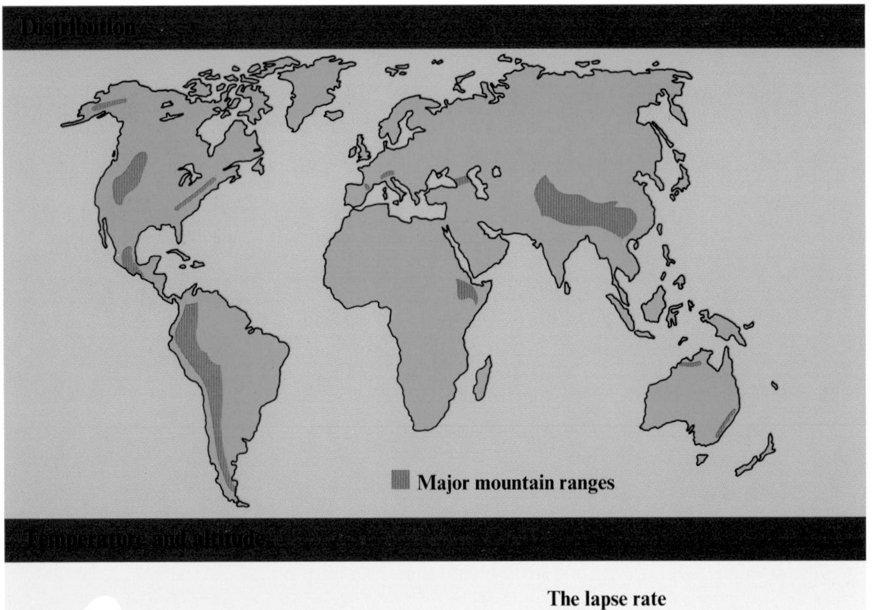

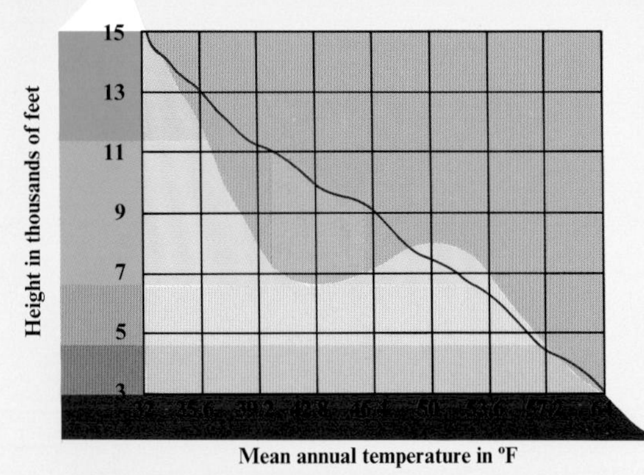

The lapse rate
Ascending a mountain, temperature decreases by 0.7°F in every 200 feet. But it may stay warmer in sheltered valleys or on sun-facing slopes

Alpine
Temperate coniferous forest
Temperate broad-leaved forest
Scrub
Sheltered valley

The Himalayas, with temperate forest trees on slopes above 5,000 feet and only snow and bare rock above the snowline at 14,500 feet

in shadow. There may be a difference, too, between the side facing into the prevailing wind, which is invariably wet and windy, and the lee side, which is much drier. Even the steepness or the stability of the slope can affect the ecological conditions.

The mountain top

The upper slopes of the mountain are rugged and exposed, and only the hardiest plants and animals survive. Just as there are fewer species at the highest latitudes, so the harsh conditions restrict the diversity of species at high altitudes.

Mountain plants usually appear in patches where soil builds up in crevices among the rocks and screes. They tend to be very small because the rarefied air increases transpiration. They also have to adapt to the often bitterly cold conditions – the alpine flower Edelweiss has hairs instead of leaves to cut heat loss. In many ways, the high mountain is like the arctic tundra, and plants are similarly simple and small, but mountain plants may have to cope with much warmer summers, because the high angle of the sun can produce very hot days.

The summer growing season is short, which is why many mountain plants are perennials, or reproduce by sending out rhizomes. In spring, mountain meadows often burst into a riot of color, with thick carpets of alpine blooms – brilliant blue gentian, stonecrops, campions, and cinquefoils – competing for the attention of insects to pollinate them.

Mountain creatures

Only a few animals are tough enough to survive the rigors of mountain summits. A few small mammals, such as the marmot in temperate regions, and the viscacha in the Andes, survive by sheltering in burrows and hibernating during the coldest periods of the year. Larger animals tend to have thick coats to keep warm. They are also agile and sure-footed like the Siberian ibex and the chamois, to reach the sparse grazing on the cliff ledges. Often they will move down to the lower slopes to avoid the worst of the winter. Owing to the scarcity of animals, large predators are even scarcer: the rare and beautiful snow leopard of the Himalayas is one of the few that still exist.

Birds can reach the very summit of the highest mountains by soaring and gliding on swirling mountain winds. Choughs have been reported at the summit of Mount Everest – the highest altitude any vertebrate has ever been seen. But the winds are fearsome enough to put off all but the strongest fliers, among them the giant birds of prey which are the most spectacular inhabitants of the high mountains. They include the great raptors, including the eagle and the mountain falcons, and carrion-feeders such as the Andean condor and the lammergaier of the European Alps. They survive mainly by feeding on small birds that live at high levels such as the snow finch, ptarmigan, and chough or that visit from time to time, but sometimes they will prey on larger mammals. The smaller birds feed on seeds and insects, which are surprisingly abundant. Many insects are blown up the mountain by icy winds and provide a rich store of frozen food not only for birds but also for other insects.

Cushion-shaped *Silene acaulis*

Sheltering
Draba oligosperma

HIGH FLIERS
Very few birds are able, or choose, to fly at high altitudes. This is not just because of the rarefied air, but because of the chill factor the winds bring to conditions that are already deathly cold. In these circumstances, most high-altitude birds such as the tinamou (16D) keep close to the ground, and may even burrow to keep warm at night. Yet there are exceptions. With its 10-feet wingspan the Andean condor (19C) is the largest of the vulturelike birds of the American continent. It occurs throughout the Andes but it is nowhere very common. Like other vultures it feeds on the carcasses of dead animals, although it has been accused of preying on living domestic stock.

The condor nests on cliffs up to 15,000 feet high. It ranges widely over the mountains and the adjacent plains and deserts for food, soaring for many miles at altitudes of up to 23,000 feet without once flapping its massive wings. It usually occurs only singly or in pairs. On occasions, however, condors may congregate at a carcass to feed. They nest every second year and the single white egg is laid in September or October. The young bird is looked after by both parents for at least a year after hatching.

PLANT ADAPTATIONS
The upper zones of the Andes have alpine meadows of grasses and herbaceous plants, called *paramos*. In southern regions especially this merges into *puna* grassland: a steppelike landscape of rough grass, herbs, and cacti.

The plants at the highest altitudes are adapted in several ways to the extremes of climate that they have to withstand. Always perennial in order to avoid the problems of regular seed germination in the difficult climatic conditions, they may have any of a number of mechanisms to resist freezing – such as hairy leaves or a high content of oils – and are also almost invariably low-growing to keep out of the swirling winds. They very often have a tussocky or cushion form that further reduces wind resistance, as well as an extensive root system which anchors them against the gusts that would sweep away less well-secured plants. Some species are highly specialized at living in crevices or cracks in rocks, or in the gaps between rubble, which provides yet further protection from the wind.

ALTITUDINAL MIGRATION

As the snows and low temperatures of winter begin to arrive on the higher pastures and forests of the mountains, the food supply of the grazing animals is reduced and they either migrate downward to lower altitudes or hibernate deep in the scree and boulder fields. As the grazing animals move downward, the larger, more active predators also migrate, following their prey, as do the smaller predators and scavengers such as the mountain fox (7F). In the Andes, the largest carnivore is the puma or mountain lion (11D). These agile cats follow both mountain sheep and the members of the camel family down the mountains to the forest and lower slopes. As well as hunting, they take carrion whenever animals are killed by winter cold and starvation. Deer and small mammals are also eaten, and pumas have been known to travel up to 50 miles a day in search of food. As the pumas move into the lowlands they may compete with, and fight, the resident jaguars that rarely leave their forest territory.

freeze in places. During the breeding season, the nests of all the species are often found on mounds of salt crystals formed by the evaporation of the lake water.

At present the total world population of Chilean flamingoes is about 300,000, and of the Andean flamingoes about 100,000. The James's flamingo is rarest of all with only 15,000 birds recorded in the wild.

Another common water bird is the torrent duck, which inhabits the fast-flowing mountain rivers and streams. This duck is especially adapted to its turbulent environment, with sharp claws and very powerful legs enabling it to cling to slippery rocks against the force of the current.

COLD BLOOD

The invertebrates and cold-blooded animals that reside in high mountains – the lizards and snakes, for example, and the Puna toad (15E) – are generally dark-colored, in contrast to the many warm-blooded animals which are mostly light-colored.

The dark coloration helps in the absorption of the sun's heat energy and affords some protection against ultraviolet radiation. The body temperature of lizards and other cold-blooded animals may reach as much as 36°F above the atmospheric temperature because of the dark coloration and through lying out in the sunshine (or by taking shelter when the weather is colder). *Liolaemus multiformis* (17E), a free-living lizard that inhabits the rocks at altitudes of more than 13,000 feet, is particularly skillful at regulating its basking and sheltering.

The lizards and toads of the Andes feed on a surprising diversity of insects, often blown up from the lowlands by the strong mountain winds and frozen into dormancy in the process, making them easy prey.

Temperature regulation in *L. multiformis*

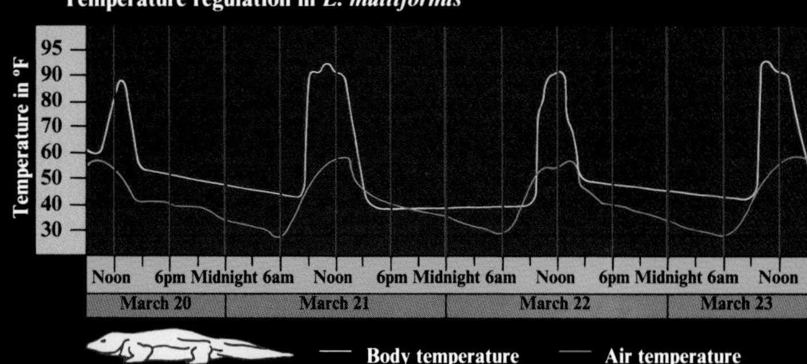

MOUNTAIN DYNAMICS

Glowing streams of red hot lava pour from the mouth of a volcano

The key to success

- Vegetation cover remains intact, with widespread root systems helping to hold the soil on steep slopes in place
- Biomass removed by herbivores does not exceed biomass produced
- The possibility of altitudinal migration – herbivores and their predators moving up and down the mountain according to the season and the severity of conditions

Forces for change

- Climate change (less snowfall, raised UV, higher temperatures)
- Herbivore population expansion, natural or by introduction, denuding the vegetation cover
- Acid precipitation and other forms of pollution
- Erosion and pollution caused by increased human recreational use (walking, skiing)
- Avalanche and other forms of erosion: vegetation/soil loss
- Volcanic eruption causing widespread destruction of flora/fauna

Mountain ecosystems are highly sensitive to any kind of disruption. At high altitudes, growth and reproduction are very slow, so the ecosystem often takes a very long time to recover from a shock, whether it is climatic, physical, or biological. The fragility varies between specific ecosystems, and between mountain ranges, but is always significant.

Steep mountain slopes are particularly sensitive to soil erosion, and the vegetation cover plays a vital role in holding them in place. Anything which disturbs the vegetation can expose the slope to the elements. The soil can then be quickly washed away – or swept away in a landslide.

In this marginal environment, the vegetation cover can be destroyed naturally by even slight variations in the climate. Plants may be damaged, for example, by a series of very cold winters – or by a succession of snow-free years – because an insulating layer of snow plays a valuable role in keeping the soil warm. Warm winters can be equally dangerous, encouraging delicate alpine plants to grow much too early in the year, only to be nipped in the bud by late frosts.

Mountain ecosystems may be especially vulnerable to changes in the atmosphere created by human activity. If global warming occurs, for example, mountain organisms could well be displaced by opportunist lowland species that have higher growth and reproduction rates. Similarly, any increase in UV radiation caused by the depletion of the ozone layer may have an especially pronounced effect in mountain regions where there are already high levels of ultraviolet radiation.

Overgrazing

Another way in which the vegetation cover can be damaged is by foraging animals. Sometimes, changes in the environment can boost populations of small mountain mammals such as voles and marmots. When this happens, they can often destroy the vegetation on which they live by overgrazing it. Soil stability may be further impaired by the simultaneous burrowing activities of an unusually large local rodent group.

Overgrazing mountain vegetation can also occur when there is a sudden increase in the normal local population of larger animals such as mountain sheep or mountain goats. This is unusual, however, because population levels tend to be limited naturally. However, farmers often override these natural controls and introduce huge numbers of domestic sheep and goats. The rise in numbers of domestic sheep on the mountainsides of the world has become a serious and increasing problem. Sheep not only transmit disease to the native wild animals but, through agricultural

Vast clouds of choking ash are blasted out by an erupting volcano

Volcanic disruption

Many mountains are volcanoes (1), and when they erupt it can have a devastating effect on the local ecosystem.

When a volcano erupts, the rapid expansion of hot gases deep inside it drives masses of red hot molten rock (lava) from its neck and blasts huge clouds of hot ash high. The lava pours down the mountainside incinerating everything in its path, while hot chunks of rock rain down all around (2). Sometimes the blast from the eruption generates a wind that can knock trees flat over a wide area. Then the ash begins to settle out of the air far around the volcano, smothering plants in a thick layer and suffocating humans and animals. Ash falling in lakes and rivers clogs the gills of fish and buries their eggs.

Yet recovery is quick. Within five months, fireweed (*Epilobium angustifolium*) and some aster species may colonize the ash, which, though lacking nitrogen, is rich in other nutrients (3). Lupins, bracken fern, blackberry, and the pearly everlasting flower (*Anaphalis* spp.) may also appear. Within a year, deer, ground squirrels and gophers may be back, adding their droppings to the nutrients and organic matter in the ash.

New growth after an eruption

The long-term effects of the eruption are greatest on the lakes, which show an increase in bacteria, in algae, and in blue-green algae as a natural result of the presence of nutrients from the ash fallout and the relative absence of fish and small crustaceans which would normally have fed on the microflora. Years later, lakes are still affected.

1

2

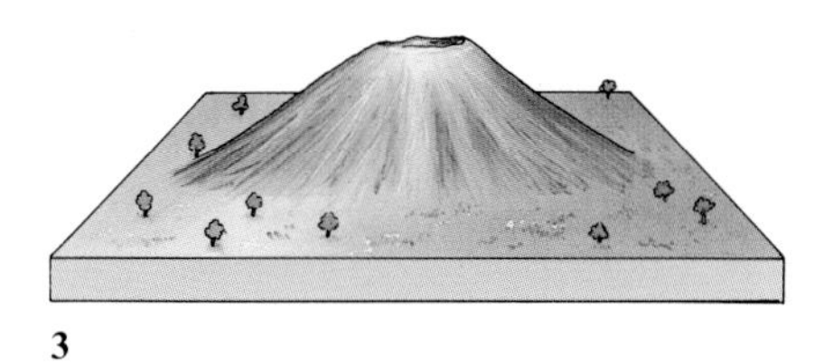

3

management, attain numbers far in excess of the grazing potential of the alpine areas. Overgrazing in this fashion has created difficulties in regions as far apart as the Andes of Bolivia and Peru and the Himalaya of Pakistan.

Skiing on acid

The impact of human activities on mountains has not been restricted to overgrazing. At high altitudes in the European Alps, the mountains of Scandinavia, and highlands elsewhere, pollution from factories often many miles away in the lowlands has brought acid rain and even acid snow. The natural nutrient cycle between plants and soils is inevitably disturbed by acid rain and, when such a disruption is intense, may even kill susceptible plant species. The effects have been particularly severe close to the tree line in mountain forests.

Perhaps less well documented is the impact of human recreational activities in many mountain areas. Visitors, sometimes in huge numbers, require accommodation, food, and fuel, all of which in turn require supply services that may ultimately lead to pollution and the disruption of ecosystems. The more shy and retiring animals – such as the larger predatory animals, and birds – may be inhibited from breeding in the face of constant disturbance by walkers and ramblers. The ground itself may suffer. Winter skiing, for example, can cause first the compaction and then the removal of the snow layers that insulate the soil, so causing damage and death to the vegetation beneath, and erosion when the surrounding snow melts or the summer rains begin.

Scree slopes

The large slopes of talus, or scree, at the foot of cliffs or steep gradients are difficult for plants to colonize. However, a few species, such as the large parsley fern, the smaller rose root, and rock campion, grow in the Alps and Scandinavian mountains. Living in the same habitat are large rodents called marmots which live in colonies, their burrows extending for many yards within the scree. Inside the burrow are nests which the marmots fill with dry grass before hibernating in October, then renew when they emerge in April. Mountains often echo to the piercing alarm whistles of marmots on watch, warning the colony of hunting eagles and foxes.

RIVER DOLPHINS

A few large river systems around the world – the Indus, the Yangtze, the Ganges, and La Plata are others – have their own indigenous species of dolphin. The Amazon River dolphin (10D), or "pink porpoise," is a freshwater species common throughout the Amazon system, from the delta to the headwaters.

A large species, it can attain 10 feet in length and weights of more than 130 lb. The species is characterized by a sturdy, toothed beak and a highly mobile head with a protuberant, fatty "melon" which can be used to focus low-frequency clicks produced in the dolphin's air passages. The ear openings are large, and the jawbone contains a fatty channel which also conducts sound well to the inner ear, for echolocation – used by most dolphins for navigation. In the Amazon River dolphin it may also represent an adaptation to life under a dense, dark forest canopy or in turbid waters. The eyes are small, though fully functional.

The most striking feature of the dolphin is its color. Young animals are bluish dark gray. As they mature, a reddish color spreads from the underside and elderly individuals may actually become completely pink.

In comparison with its marine cousins, the Amazon River dolphin swims relatively slowly. It usually lives in pairs. The diet is mainly fish – including the notorious piranha – with some crabs and prawns.

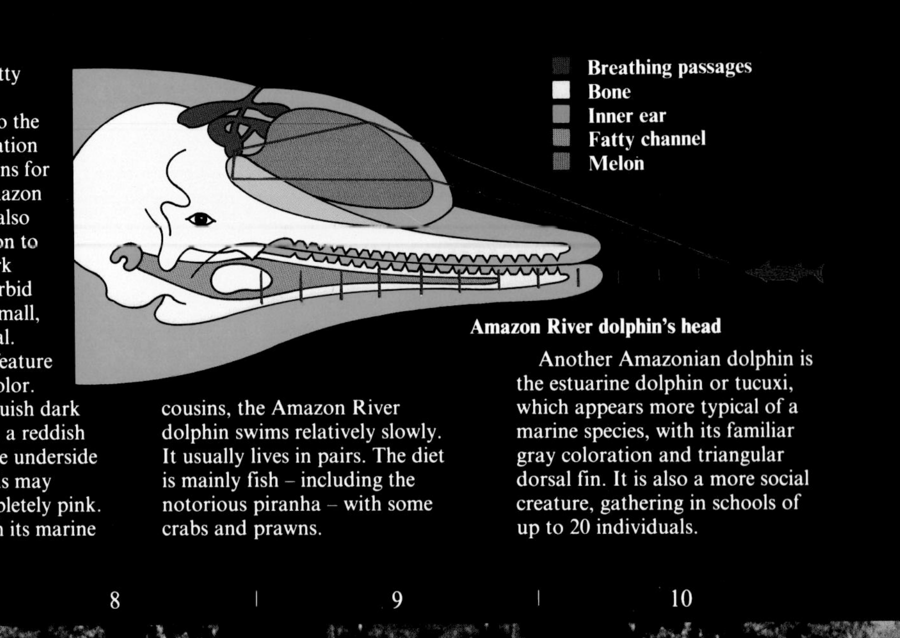

Amazon River dolphin's head

Another Amazonian dolphin is the estuarine dolphin or tucuxi, which appears more typical of a marine species, with its familiar gray coloration and triangular dorsal fin. It is also a more social creature, gathering in schools of up to 20 individuals.

6 | 7 | 8 | 9 | 10

6 | 7 | 8 | 9 | 10

RIVERSIDE INSECTS

Around 8,000 species of insects have already been recorded in the Amazon basin, and insect life plays a major role in the nutrition of the river's ecosystem. Many fish have developed specialized feeding apparatus – such as the forward-pointing sensory barbules of the arowhana (12B) – to detect and capture insects and spiders that fall on to the water's surface. Many fish follow the annual floodwater in through the forest, dining on the innumerable terrestrial insects now drowned. Dense flocks of butterflies congregate on the exposed mudbanks of the river as the floodwater recedes (1D), attracted by salt within the mud.

TEMPORARY POOLS

Killifish (15B) occur throughout the Amazon region, especially in temporary pools left behind after flooding. The life-cycle of these fish has evolved in order to carry out the hatch-grow-reproduce-die regimen within nine months – the typical total duration of the temporary pools. The eggs are buried in pool sediments, where they endure a long period of dry incubation as the ponds evaporate; they hatch when the rains or the floods return. The rewards of such a lifestyle are that the pools are relatively free from predators, and that the invertebrate food is plentiful. A temporary pool may be less than 8 inches deep.

The Amazon is the largest river basin in the world, with a total catchment area exceeding 2.7 million sq miles, or nearly half the landmass of South America. Its principal headstreams rise in the Andes, only 100 miles from the Pacific, but its 50 mile-wide estuary empties into the Atlantic some 4,000 miles away across the continent.

About 5 million years ago the Amazon Basin was probabl a vast, enclosed freshwater sea, before breaches appeared i its eastern rim and drained its waters into the Atlantic, to form the river system that we know today. So great was th scouring effect of this drainage that the beds of many of th Amazon's deeper channels lie below sea-level. This aweson waterway is home to the greatest diversity of life found in any river: there are 20 different species of stingray alone.

1 2 4 5

A B C D

1 2 3 4 5

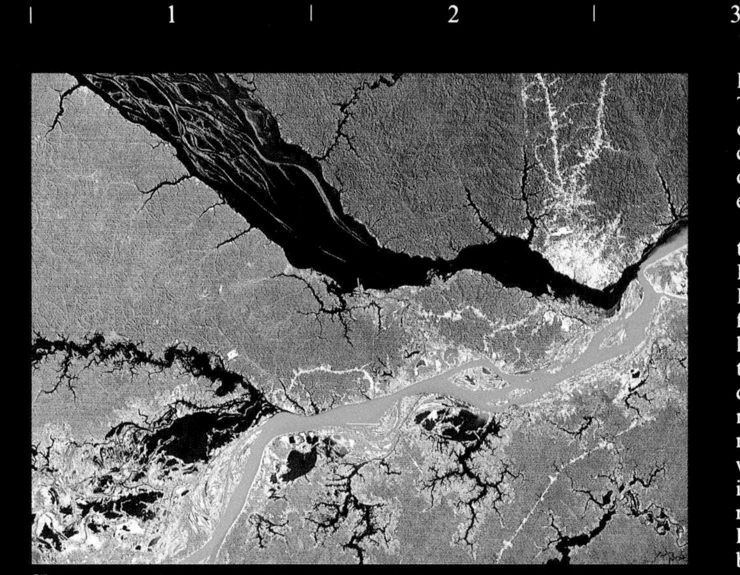

Clearwater and blackwater tributaries of the Amazon, showing the clear boundary between their waters after they have joined

RIVER-WATER DIVERSITY

The headwaters and tributaries of the Amazon system display a complex array of waters of different chemical characteristics, each suiting different forms of life.

Clearwater rivers arise from the hard, insoluble rocks of the Brazilian and Guiana Highlands. Blackwater rivers run mainly from the Amazonian rain forest lowlands, which occupy much of the western basin. They are very dark in color because they run off nutrient-poor soil, which limits microbial activity so that the waters carry a high proportion of incompletely-decomposed organic matter from the rain forest. Paradoxically, this makes the blackwater rivers the most chemically pure natural waters in the world – some contain lower solute concentrations than ordinary distilled water.

As the Amazon empties into the Atlantic, it is a whitewater river – turbid with heavy sediment. The diversity of water types encourages a large variety of specially adapted fish, such as the discus fish (10E) which, living in nutrient-poor waters, feeds its young with slime produced on its flanks. More than 1,500 fish have been described, and an estimated 1,000 species await discovery. Where different waters meet, differently adapted species may be only yards apart, yet separated as effectively as by a glass wall.

RIVERS AND LAKES

Just a fraction of a percent of the water on the Earth's surface is freshwater. But wherever freshwater collects it becomes a rich habitat. Freshwater habitats are found nearly everywhere in the world, from the equator to the poles. Only in deserts, where water is scarce, and in the polar regions where water freezes to ice is freshwater life entirely absent.

There are flowing water habitats, which can vary from an icy mountain brook to a vast tropical river like the Amazon. And there are still water habitats, from tiny rainpools in the forks of tropical rain forest trees to giant lakes like Lake Baikal in Siberia, many thousands of feet deep. Each kind has its own range of plants and animals.

As they flow from source to sea, most rivers change dramatically. A river typically starts high in the mountains as rainfall and water from underground gathers together to form a rill. In its upper reaches, a stream is usually cool and crystal clear, tumbling over boulders and gravel. Sometimes, the water plunges over rapids and waterfalls; sometimes it idles through pools. Very few plants can live in the shifting stony bed of the stream, but mosses, liverworts, and ferns often adorn the banks. Animals survive this turbulent environment by sheltering in crevices like the freshwater shrimp, or clinging to rocks with claws like the stonefly and mayfly or with suckers like leeches. Insects are typically "suspension feeders" like the blackfly and caddis fly larvae which sieve food from the water as it sweeps swiftly past. Fishes must be strong swimmers with streamlined bodies like the trout and stone loach, which appreciate all the oxygen in this cold, bubbling water. Further downstream, the river falls less steeply and rapid sections become less frequent. Here the flow regime suits fish like the grayling.

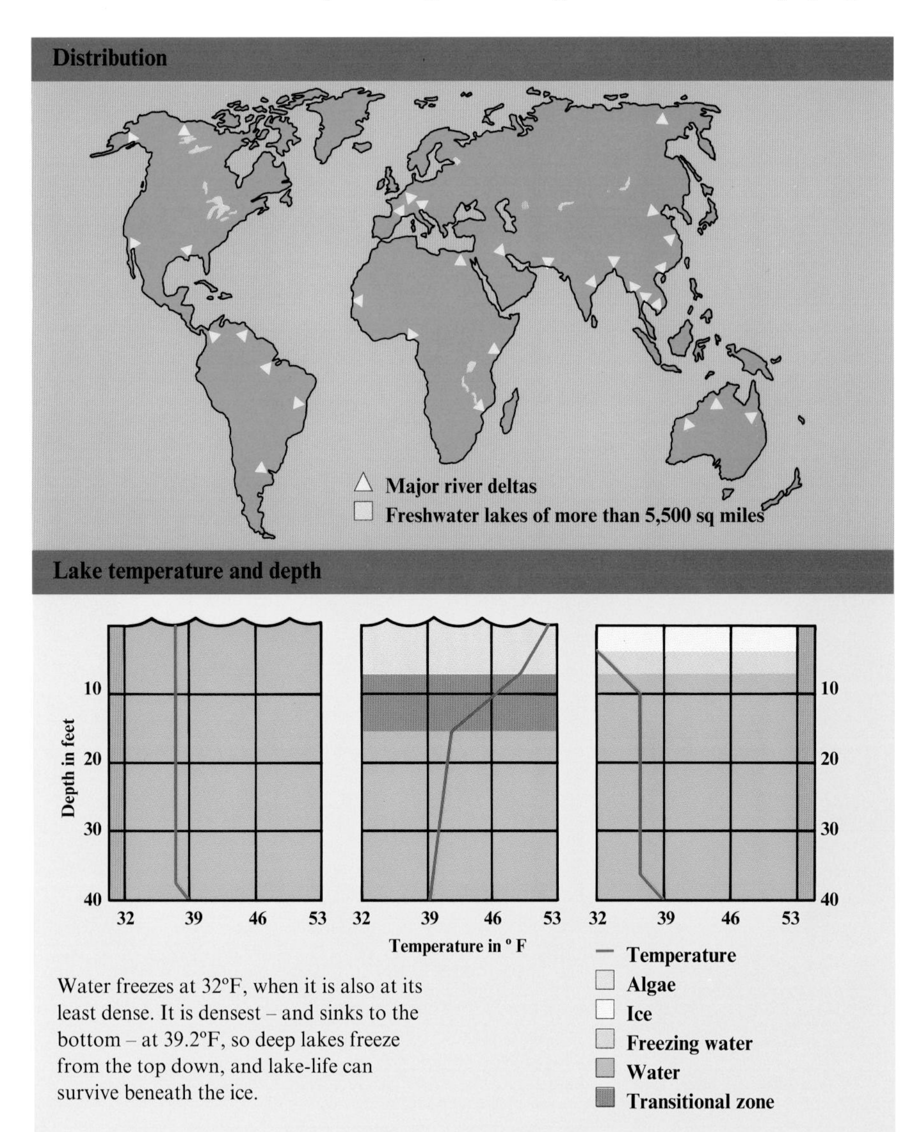

Water freezes at 32°F, when it is also at its least dense. It is densest – and sinks to the bottom – at 39.2°F, so deep lakes freeze from the top down, and lake-life can survive beneath the ice.

As the river emerges from the foothills of the mountain region, the gradient becomes much more gentle, it begins to wind a little, and silt and sand are deposited, making the bed much smoother. More plants appear, putting down roots in the sediment, where mollusks and worms burrow. Insects such as the alderflies, mayflies, and beetles feed on the plants both as adults and larvae. The plentiful insects, in turn, draw birds such as dippers to the stream. The gentler flow and abundant food encourage a richer variety of fish in this section, including barbel and eels.

As it nears the sea, the river is deep and smooth-flowing, meandering across broad flatlands which it often floods. Here the river life is at its richest. Common bream graze on the streambed, eating insect larvae, worms, and mollusks. Schools of fish, such as bleak, feed on floating plankton or break surface to snap up flies. Predatory perch and pike lurk in the thick, fringing reeds and rushes, where all kinds of insects and birds such as ducks, grebes, and herons live. The riverbed is alive with burrowing creatures such as freshwater mussels, while mammals such as voles and otters live on the bank. Oxygen levels are often low near the bed, and many burrowers have red oxygen-carrying hemoglobin pigments like the bloodworm to make the most of what there is. In the tropics, the fish population may include voracious carnivores such as the piranha. Here, rivers often provide the easiest path through the dense forest, and as well as water snakes and amphibians such as caimans and crocodiles, there may be hippopotamuses and, along the banks, elephants, and tapirs.

Lakes

Lakes, pools, and ponds form where water collects in a dip in the ground. Such dips form in all kinds of ways, and lakes occur in many different situations. Many form in hollows carved out by glaciers during the Ice Ages, like the Alpine lakes, and the Great Lakes of North America. Some of the largest and deepest are created by the movement of the immense plates that make up the Earth's surface. The world's deepest lake, Lake Baikal in Siberia, is just such a lake, as are the massive lakes that form a chain in the Rift Valley in eastern Africa. Most lakes and pools are natural, but in recent years, many reservoirs have been constructed to supply water to growing cities.

The natural life of a lake depends on the character of the water – in particular, its chemical content and the frequency with which water is replaced. These in turn depend on the size and catchment area of the lake, and the nature of the rocks and soil. The water in a large lake – or one with a small catchment area – remains in the lake basin for a long time, increasing the delay in recycling nutrients within the lake. The quantity of nutrients flowing through the water also exerts a powerful influence on oxygen levels and lake productivity.

"Eutrophic" lakes are rich in nutrients (and therefore rich in animal and plant life). They are shallow and water flows into them from non-acid soils, bringing calcium, nitrates, and phosphates. Deep, clear, "oligotrophic" lakes form in areas less rich in nutrients, and have a much less abundant but often diverse plant and animal life. "Dystrophic" lakes exist only in areas with very acid soil, such as heathlands, and plant and animal life is much more limited. Nevertheless, different plants have evolved to exploit waters in which different levels of nutrients are available. Whereas some such as the red pondweed (*Potamogeton alpinus*) are adapted to nutrient-poor waters, plants such as the yellow waterlily (*Nuphar lutea*) are adapted to nutrient-rich waters. The duckweeds (*Lemna* spp.) occur in waters of widely different nutrient availability thanks to their capacity for growing roots long enough only to suit the nutrients that are available.

Phytoplankton – typically green algae – are abundant in nutrient-rich lakes in spring and summer. Indeed, eutrophication – an excess of nutrients from sewage, for example – can cause algal bloom which virtually chokes the lake. The algae provide food for zooplankton, protozoans, and shellfish which in turn provide food for fish such as roach, tench, and rudd. Pike and perch feed on the grazing fish. Where the water is shallow, waterlilies, duckweed, and pondweed float on the surface, with their roots in the lakebed, while reeds and rushes emerge on the lake edge. Among these plants, mollusks, worms, and insects live in the mud on the lake bed while the water holds plankton, shellfish, and fish. Moorhens and coots, voles, and numerous insects swim at the surface, while birds such as reed warblers nest among the reeds.

THE MANATEE
A large, gentle mammal, the Amazonian manatee (8C), basks in the warm waters of the Amazon River system, grazing on aquatic vegetation. Like the dolphins it is related to a sea-dwelling species, which lives off the coast between the northwestern United States and northern South America. Manatees are social animals and congregate in herds that migrate between the main rivers and tributaries and connected lakes. The shortsighted group members communicate by muzzle to muzzle touching and, when alarmed, by chirping. Breeding typically takes place every four years: sexual maturity is reached after seven years. The gentle pace reflects the manatees' peaceful existence in the waters of the Amazon – but has proved costly over the last century. Human hunting of the manatees has brought them near extinction.

16 | 18 | 19 | 20

A — B — C — D

16 | 17 | 18 | 19 | 20

FLOODS IN THE CANOPY
The huge catchment area and the flattish river basin of the Amazon come together to produce temporary but dramatic fluctuations in river level. When the river system is at low flow, the main streams and major tributaries are constrained within discernible banks. But from January to May the river can rise by as much as 30 feet, and the streams can become hundreds of miles wide. Areas of canopy forest are inundated and opened up for fish to feed and breed in. Spider monkeys (16A) and other animals can be caught out by the flood and may be marooned in the branches of trees. The vast, temporarily flooded area – known locally as the *varzea* – is not swampy during the dry seasons because it lies well above the normal level of the river.

Many species of tree are adapted to this periodic and extensive inundation, together forming a temporary swamp forest, notable among them the majestic kapok (or silk cotton) tree with its tall, erect, buttressed stem and spreading canopy.

For the rivers to descend back to their dry-weather level may take only a matter of days. Many fish are stranded in pools and on dry land, providing food for forest animals and nutrients for the vegetation. This intricate, supportive interaction between forest and river ecosystems is significant for sustaining the entire Amazon.

Many non-aquatic animals make a living from the river. The 40 feet-long anaconda (4D) is an excellent swimmer, and may even tackle and kill young caimans (3D). It does not, however, actually crush its victims to death, but merely asphyxiates them.

The largest of the several species of otter is the giant otter (2D), otherwise known as the saro or flat-tailed otter. It can attain a length of 6 feet and weigh up to 90 lb. Very rare, the giant otter is confined to slow-flowing waters. Like other otters, it feeds on small animals in the water, including fish and small aquatic mammals. But unlike most other otters, the giant otter has a flattened, ridged tail and is covered by dense brown fur. These are intelligent and playful creatures, and form close family groups whose members call to one another constantly in order to maintain contact in densely vegetated areas. It is their calling that renders them easy to track by hunters, who have exploited them almost to the edge of extinction for their especially luxurious fur. The gradual loss of their habitat is now making their survival even more precarious.

RIVER AND SEA

There is no clear boundary where the freshwater Amazon ends and the salt sea begins, and this has major implications for the residents. The dolphins which inhabit the river probably evolved from saltwater species that penetrated far upstream. The Amazon does not form a delta at its mouth, even though it is estimated to discharge a vast 1.5 million tons of sediment into the sea every 24 hours. The reason lies in the rate and the turbulence with which the waters flow at the estuary, resulting from the huge volumes of water draining from the Amazon basin, and from unusually large tidal fluctuation. In fact, the tidal effect can be measured up to 620 miles inland: one manifestation is a tidal bore 5 to 12 feet high, which sweeps upriver as a result of the funneling influence of the riverbed.

Sediment transported down the river may not create a typical delta, but it does result in spectacular formations below the water. The layer of sediment deposited at the river-mouth is more than 1¼ miles thick, ridged by dunes up to 35 feet high.

The river-mouth frequently floods: perhaps a fifth of all the freshwater that drains into the sea worldwide flows from the Amazon. Such is the rate and quantity of flow that the sea is diluted up to 100 miles out from the river-mouth. Fresh- and saltwater generally mix at river-mouths, but so great is the force of water flowing from the Amazon that the lighter freshwater forms a layer over the denser seawater, and mixing occurs only slowly in the zone where these meet. Freshwater can be scooped from the sea surface as far as 40 miles offshore.

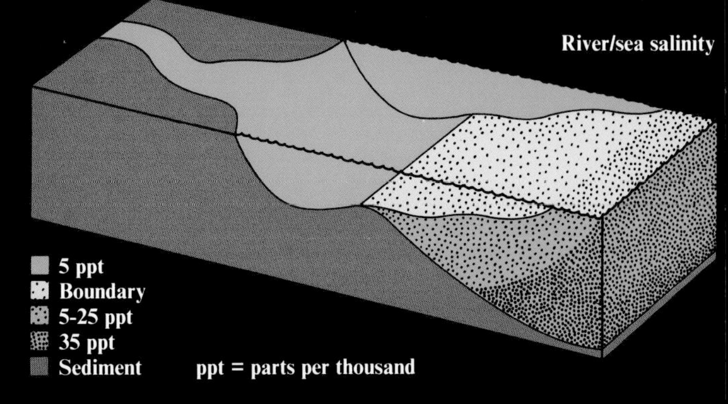

11 | 12 | 13 | 14 | 15

11 | 12 | 13 | 14 | 15

BODY SHAPE

Fish like the stingray (15E), or any of the numerous species of Amazonian catfish (12E), which scavenge plant and animal matter from the river bed, typically have flattened undersides with a wing-shaped upper surface to help them hug the river bed. Fish such as the cardinal tetra (11D) are adapted to an active mid water-life. They are streamlined, agile and forever on the move. Surface feeders like the arowhana (12B) are often slim to allow them to maintain their position in the current as they wait for prey to drop onto the water.

The much feared piranha (7E) have large, razor-sharp teeth and blunt, powerful bodies, ideal for acceleration and maneuverability in the melee of a feeding school. Renowned for being able to strip a tapir (17C) carcass to the bone in less than three minutes, piranha go into a frenzy at the scent of blood, but otherwise will often allow large animals to swim among them unmolested.

The skin of the knife fish (12C) has some cells which generate electrical impulses, and others which detect them. By holding its long body tense and straight – and swimming forward or backward by wriggling its fins – it can generate an electrical field and navigate according to how this field is disturbed by surrounding objects.

RIVER AND LAKE DYNAMICS

The key to success

- Animal-plant interactions on lake- or river-bed
- Animal-plant interactions in the plankton
- Adaptations to exploit individual characteristics of the water body (such as certain types of algae that live in sulfur lakes)

Forces for change

- The presence of oxygen-depleting material in the water
- Suspended solids in the water, often caused by physical disturbance such as leisure activities
- The use of waterways for the dumping of sewage
- Pesticide runoff from agricultural land
- Oxygen depletion by sudden excessive plant-growth
- Low rainfall and/or excessive abstraction of water
- Faulty large-scale damming schemes
- The natural formation of lakes, as a result of rock-faulting or river-looping

In their lower reaches rivers often form intricate looping meanders

Most rivers and lakes are inhabited by a wide variety of waterplants. Some (such as pondweeds) are submerged, some (such as waterlilies) have their leaves floating on the surface, and some (such as rushes) have erect shoots that stand up out of the water. On and above the water, the plants maintain themselves by photosynthesis, but below the waterline their surfaces provide a home for large colonies of algae which tend to take up residence wherever there is enough light for their own photosynthesis.

The algae themselves are food for grazing animals such as water snails and the aquatic larvae of many types of insect. The waterplant shoots are grazed less extensively during the growing season, but their structure softens as they begin to die back during the autumn. As the plant breaks down into fine organic particles, it becomes available for a much wider range of animals to graze on. The particles may be extracted from the water by filter-feeding animals such as the larvae of the blackfly, *Simulium,* or may sink into the sediment where they are browsed on by bed-crawling animals such as the water skater, *Asellus*.

Plankton

The algae form one part of the complex community that is known altogether as plankton. The phytoplankton (microscopic plants) includes many different types, ranging from small, unicellular algae through to large, filamentous varieties. There are, similarly, many different sizes of zooplankton (microscopic animals), ranging from the microscopic rotifers through the larger, herbivorous water fleas, right up to "giant" predatory water fleas such as *Leptodora* which can often grow up to a third of an inch long.

Not all the zooplankton are equally efficient at grazing the algae. The smallest, such as the tiniest rotifers, may be able to eat only bacteria and the most minute algal cells. The larger rotifers and smaller water fleas, which can easily digest diatoms and small green algae, cannot consume large, filamentous algae, and this is the food of the more sizeable water fleas and similarly efficient grazers. The largest members of the zooplankton are also predatory upon smaller planktonic animals.

The next stages of the food chain continue the same theme: small fish consume the largest of the zooplankton, and larger fish consume the smaller fish. The great diversity of such living organisms in this complex food chain provides the ecosystem with the resilience to remain stable when conditions change, as they may do in a seasonal cycle.

At the same time, the ecosystem's stability is to some extent self-regulating

In an oxbow lake, the flowing-water ecosystem of the river is replaced by a lake ecosystem

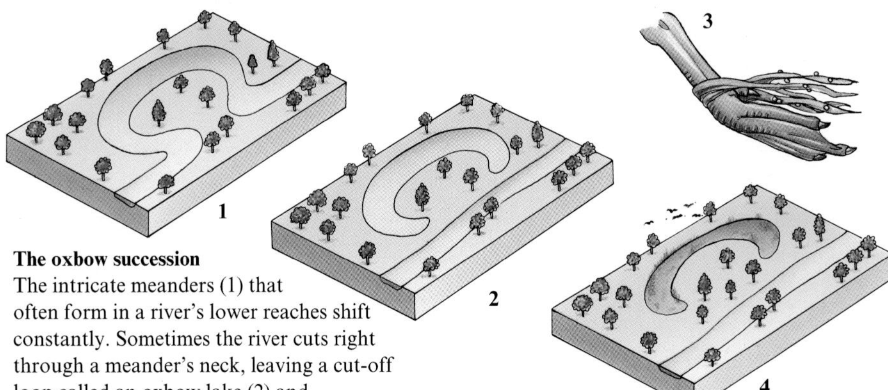

The oxbow succession
The intricate meanders (1) that often form in a river's lower reaches shift constantly. Sometimes the river cuts right through a meander's neck, leaving a cut-off loop called an oxbow lake (2) and transforming the ecological regime here. Plants adapted to flowing water are overgrown by algae and ousted by lake plants, while river fish dependent on high oxygen levels give way to lake fish, introduced as eggs on plants clinging to ducks' feet (3). As the lake silts up, released nutrients produce a wealth of plankton. Submerged waterplants are replaced first by species with floating leaves (waterlilies) and then by emergent species (reeds). Trees tolerant of waterlogging (willow) encroach, and invertebrate and fish communities change as the food and oxygen supply diminishes. Finally, aquatic and wetland vegetation chokes the water, forming a bog (4).

thanks to the web of mutual interdependences. If fish predation is low, for example, large carnivorous zooplankton can survive and prey on the large zooplanktonic herbivores, keeping their numbers down and allowing smaller zooplankton to flourish in the absence of competition. These small zooplankton in turn graze upon the smaller algae, so that the proportion of larger algal species in the ecosystem rises or submerged waterplants proliferate.

To this imbalance the ecosystem can respond correctively. Fish fry multiply with all this food, decimating the population of predatory zooplankton. In turn, the larger herbivorous zooplankton flourish as predation on them is reduced and they are able to exploit the abundant large algal species in the water. The smaller zooplanktonic grazers suffer as the large water-fleas graze more effectively, and so the proportion of smaller algae tends to rise.

Clear and murky waters

The water of some lakes seems always to remain clear, whereas water in other lakes tends to be murky. In still waters, apparent murkiness is often a result of dense growths of plankton, often in response to a rich supply of mineral nutrients. Murky water generally means that the ecosystem is dominated by plankton, and clear water most often indicates that the ecosystem is dominated instead by the waterplants.

The natural stability of ecosystems means that a murky lake does not turn into a clear one, nor clear into murky. Turbid water, after all, does not favor the predation of plankton-grazing fish by carnivorous fish such as pike, which rely on direct vision of the prey. Conversely, where large stands of submerged water plants develop, the vegetation cuts off nutrients and light, suppressing phytoplankton. The resulting clear water, and the ambush sites provided by the vegetation, mean that carnivorous fish prey more effectively on plankton-eating fish. Larger cladocerans flourish and graze on what algae there are, keeping algal populations at a constantly low level.

Disruptions to the ecosystem

Human activities often damage freshwater ecosystems by producing fluctuations in conditions greater or more persistent than those which occur in nature, or by introducing toxic matter. The problems that most frequently arise are those caused by pollution, by physical disturbance, and by water abstraction.

Of the sources of water pollution, the five most commonly encountered are: the presence of oxygen-depleting material; the presence of suspended solids which increase turbidity, clog fish and invertebrate gills, and again reduce oxygen levels; the presence of ammonia, produced by the decomposition of nitrogen-containing compounds (as in sewage); the introduction of mineral nutrients, causing a bloom of plant growth and again reducing oxygen levels; and the introduction of toxic synthetic chemicals (such as pesticides).

Human activities at the waterside, in or on the water itself, cause considerable physical disturbance. Water sports are disruptive to the animals and fish in the water, and to fish-catching birds. Moreover, the increased turbulence generated by boating, jet-skiing, and other water sports, frequently gives rise to erosion of bankside and littoral vegetation, and may also damage other valuable habitat features such as mudbanks or fallen trees. Sediment is also brought into suspension by the turbulence, clogging invertebrate and fish gills and reducing oxygen. Habitat richness is reduced, and so, therefore, is biological diversity.

COLD ADAPTATIONS

Deserts experience extremes of both heat and cold, because the atmosphere that overlies them contains little humidity: ambient temperatures are not moderated. The Great Basin of Nevada may go from near 104ºF in the day right down to freezing at night. The desert lizard takes refuge in rock crevices at nighttime as well as in the hottest part of the day, because although the air may be freezing in the hours of darkness, the rocks continue to release heat absorbed during the day.

DESERT REPTILES

The western diamondback rattlesnake is a pit-viper (11D) that is well camouflaged amid desert terrain. The infrared pits beneath its eyes allow it to detect warm-blooded rodents during its nocturnal hunting rounds.

One of only two venomous lizards in the world is the colorful beaded gila monster (19D), which inhabits the hot deserts of the United States and Mexico and, like the viper, eats rodents, though it is too slow to catch active adults. The long legs and leaping ability of the kangaroo rat (18D) allow it to escape from most of its venomous predators. Leaping also minimizes its contact with the hot sand, an important and common desert adaptation. Scampering pinnicate beetles lift themselves high off the ground to run swiftly over the desert sands. One of the most specialized methods of desert locomotion is sidewinding, used by vipers and rattlesnakes, in which the body is thrown forward in successive, muscular coils.

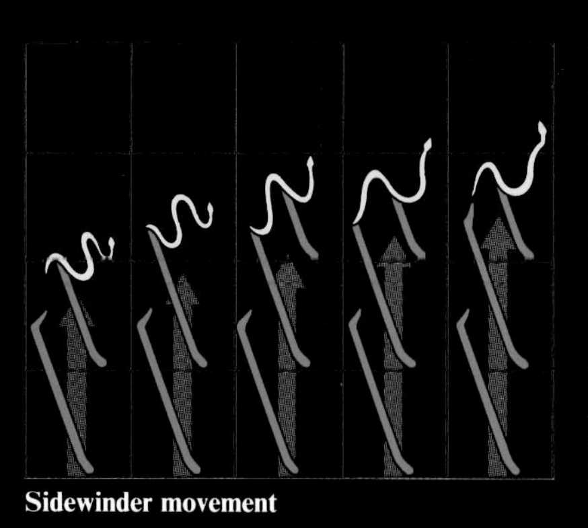

Sidewinder movement

6 | 7 | 8 | 9 | 10

6 | 7 | 8 | 9 | 10

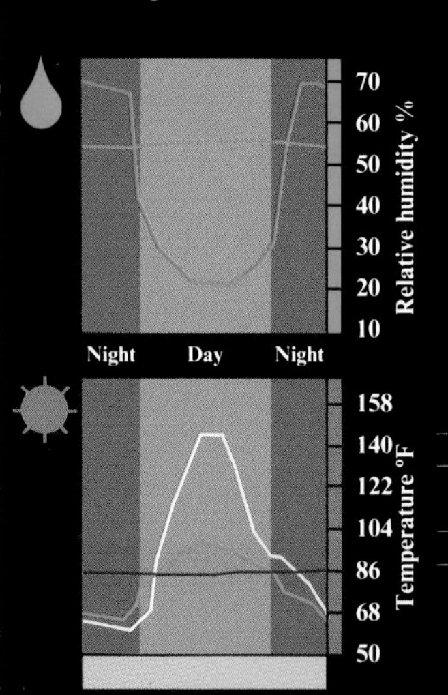

Burrow humidity
Air humidity
Burrow temperature
Air temperature
Sand surface temperature

HEAT ADAPTATION

In the desert, only about 10 percent of the Sun's radiation is screened by clouds. Animals that are active by day are accordingly usually light in coloration – often a tan or yellowish rufous in furry animals such as the large-eared desert kit fox (5D) of the western American deserts. In a notable case of parallel evolution, the kit fox has a virtual twin in the fennec fox of North Africa.

The attenuation of parts of the body, like the fox's ears, into large radiative surfaces, is a well-known adaptation for heat dissipation. In addition, most mammals possess sweat glands that allow surface evaporation to remove excess body heat. Wild horses (7C) are especially well adapted for this.

Heat regulation in reptiles requires considerable behavioral adaptation. Many desert lizards, such as the chuckwalla (17B), slide into crevices between rocks in order to avoid the scorching heat of the sun. Others may burrow, creating for themselves a small, more stable and much more comfortable environment.

Rocks themselves may contain cavities and niches, which are then available as watering troughs, especially to small organisms such as insects and spiders, to birds, and to some of the smaller vertebrate mammals.

The Great American Desert is some 500,000 sq miles in area, occupying an expanse of rugged territory in central eastern North America. The conditions are mostly maintained by the surrounding rain shadows of the Sierra Nevada and Cascade ranges to the west and the Rocky Mountains to the east. The western mountains are the more important, because the prevailing winds sweep in from the Pacific, only to be drained of their moisture as they are forced to rise as they pass over the coastal ranges.

THE BIRDS OF THE DESERT

The Great Basin of Nevada is home to the majestic golden eagle (14B). The ability to fly tirelessly represents a considerable advantage over terrestrial animals' means of locomotion through the desert. Many birds that nest in forests or grasslands on the edges of the desert – such as the Californian quail (2D) – fly long distances to utilize remote and sparsely distributed food resources. Other birds have evolved special adaptations to permit them to remain permanently in the desert.

The gila woodpecker (20A), like several desert birds, makes its home in cacti. Birds mostly have higher body temperatures than mammals anyway, so the desert heat is less of a problem, but to minimize the uptake of radiant heat one common habit is to compress their feathers close to the body. However, the most common method that the birds have evolved to keep cool when resting is to flutter a patch of skin at the front of the neck.

INVERTEBRATES

The desert is a natural home for insects, providing a wide variety of niches that accord well with the insects' various and usually very distinct life stages (one stage at least including the possibility of flight). Ants are particularly successful, and harvester ants (4D) may strip 480 million plants per acre of vegetation on sagebrush steppe, transporting seeds and leaves to their mounds as food stores. But the desert ants are themselves prey to many lizards. The food chains and webs of deserts are generally simple, in comparison with other ecosystems of the world, in which a greater diversity of species is assembled.

Desert arachnids, such as spiders and scorpions, thrive in their habitat. Hairy tarantulas, black widow spiders, and trap-door spiders all possess ingeniously specialized adaptations to the environment.

Arachnids and insects frequently burrow into the desert sand in order to avoid the extremes of heat or cold. Most desert insects and spiders are of light coloration, but a surprisingly large proportion are in fact black instead. Darker colors absorb heat more readily, but also radiate it very well, and this apparently odd adaptation is thought to allow the surface re-radiation of absorbed heat.

Black widow spiders (18D) inhabit the most remote, dry rocky ranges of Nevada, where they may find minute quantities of water in crevices or among pockets within the volcanic rocks.

Desert spiders, like many other creatures, may remain dormant for long periods of time, being re-animated only when the opportunity arises either to eat or to reproduce. Scorpions and the related whip scorpions (12D) are highly adapted desert-dwellers capable of long periods of suspended activity. Scorpions are also distinguished by carrying their young on their backs, which minimizes the dangers of predation as well as exposure to the hot desert sands.

DESERTS

Over a third of the world's land surface is so inhospitable to life that it can be called desert or semi-desert. Certain deserts, such as the great Sahara desert, are some of the hottest places in the world, with scorching sunshine beating down relentlessly day after day. Others are bitterly cold; winter temperatures in the Gobi desert are often below –5°F. But all deserts are dry, though not always because it rains so little – a desert is defined as an area where more water evaporates than falls as rain. And without water, survival in the desert is very difficult for all forms of life.

Yet very few deserts are actually completely barren. Nearly everywhere life has managed to gain some kind of foothold, each plant and animal adapting in its own way to the harsh conditions, and deserts are surprisingly lively and varied biomes.

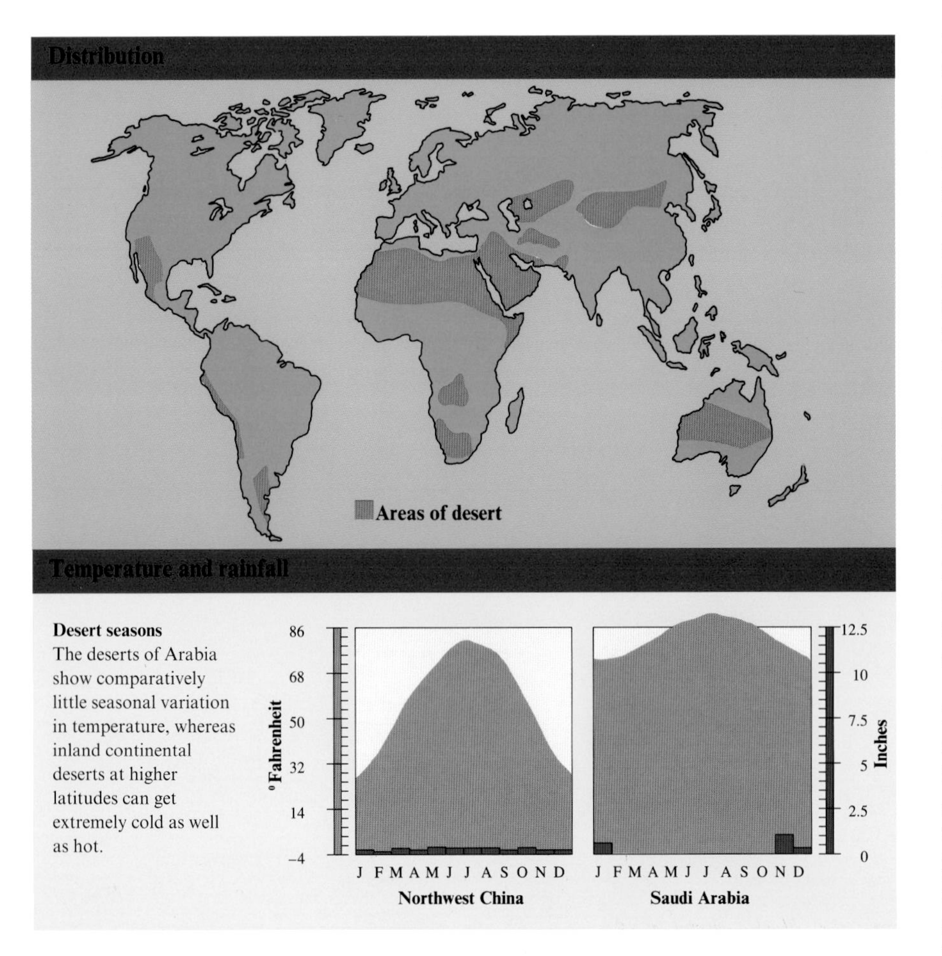

Desert seasons The deserts of Arabia show comparatively little seasonal variation in temperature, whereas inland continental deserts at higher latitudes can get extremely cold as well as hot.

Many of the world's largest, driest deserts lie within the subtropics, along the high-pressure belts north and south of the equator where the air is nearly always calm and clear. The Namib and Kalahari deserts in Africa, the Arabian desert, the Great Victoria desert of Australia, and the western coastal deserts of America are all within this zone. So too is the Sahara, the world's largest desert, made up of 3.5 million sq miles of dry rock, sand, and mountain. In these deserts, it is almost invariably very hot during the day, and the little rain that falls evaporates almost immediately. Summer temperatures in the Sudan desert can soar to over 120°F, while temperatures at Dallol in Ethiopia average 94°F all year round. But it can get surprisingly cold at night, because heat escapes into the clear skies.

Deserts also occur deep in the heart of continents, so far from the sea that moisture-laden winds never reach them – for example, the Gobi and Tibetan deserts north of the the Himalayas. Others, such as the North American deserts east of the Sierra Nevada, occur in the lee of mountains with all the moisture in the winds having fallen as rain on the windward side.

Some of these deserts can get extraordinarily hot, but where they are at high altitude, for example the Gobi, they can often be quite cold. In winter, temperatures in the Gobi can plunge to –4°F. Some scientists refer to the cold polar regions as deserts, too, because here water is locked up in ice, creating what is in effect a drought for plants and other organisms.

Warm deserts typically get less than 5 inches of rain a year, and most of that evaporates. In some deserts, years go by without a drop of rain; then when rain eventually comes, it tends to run straight off the hard-baked ground in torrents – sometimes creating raging flash floods through old streambeds or "wadis," which are often dry for years at a time.

With so little water, it is not surprising that some desert areas are completely barren of vegetation. Nonetheless, most deserts do have at least a sparse cover of plants. There is scant soil in the desert, but a few tough grasses, cacti, thorn bushes, and other spiny shrubs manage to put down roots in the sand, and around the fringes of the desert gnarled and spreading woody plants survive.

Along old water courses, there may be tall grasses, acacias, and isolated trees such as baobab trees, storing water in their bulbous trunks. Ironwood, smoketree, and palo verde grow along the courses of intermittent streams – not for the water but because their seeds need to be scraped and bruised by the churning sand of a flashflood before they can germinate.

Water-saving plants

Desert plants are adapted to minimize water-loss. Typically, they have thin, tough leaves with stomata (breathing pores) that only open in the cool of the night. Some desert grasses have folded leaves with the stomata inside. Cacti dispense with leaves and branches altogether, and instead store water in their tough succulent stems. Desert plants also tend to have long, branching root systems with which to draw water from the large area; some have tap roots, such as the tamarisk, that go deep to reach the water table far below ground; others, such as the *Larrea*, spread far just below the ground to catch every shower of rain.

A few desert plants are ephemeral, coming to life only during rainy periods, but otherwise remaining dormant as fruit or seed. In the North American deserts there are two rainy periods, one in winter and one in summer: different species flower after each period. The North American creosote bush is deciduous, putting out its leaves only during the winter rains.

The desert floor is generally bare, and the soil is thin or nonexistent. In a few places, dead leaves and twigs collect, blown there by the wind, but there is not the thick layer of organic debris common in damper areas.

Fungi and actinomycetes are the only soil organisms present in any quantity, and their activity is limited to short rainy periods. Decomposition of plant matter tends to rely on termites, which build distinctive mounds here and there. Plants therefore have to rely much more on their own resources for nutrients, and desert plants tend to retain elements such as nitrogen, phosphorus, and potassium in their stems and roots. When they are shed, these parts accumulate beneath the plant, which thus creates its own small island of fertility.

Desert creatures

Deserts are harsh environments for animals to live in. There is little food, little water, and little shelter from the blazing sun (and cold nights), so desert creatures are specially adapted to combat these problems.

Many cope with the heat of the sun by burrowing underground. The difference in temperature between the scorching surface and even a shallow burrow can be quite astonishing. Some mammals, such as naked mole rats, live underground all the time; others, such as the fennec fox of North Africa, rest in their burrows during the day and come out to hunt only at night. Rodents such as the kangaroo rats of North America, the jerboa of North Africa, and the ground squirrels of the Americas are active mostly at dawn and dusk, but shelter underground during the middle of the day when the sun is at its hottest.

Insects, scorpions, and spiders, as well as snakes and lizards, especially geckos and skinks, adapt remarkably well to desert life, but even they usually hide from the sun in burrows or beneath rocks. Lizards rest in rock crevices during the heat of the day, and come out to hunt and bask in the sun only at dusk.

Desert creatures have a variety of ways of coping with the lack of water: by not sweating, for example, or producing very concentrated urine. The addax, a large Saharan antelope, gets all its water from its food, so never needs to drink. The same is true of the kangaroo rat, which also saves water by eating its own droppings. The thorny devil, a spiny Australian lizard, soaks up water from damp sand through special scales. Camels regulate their body temperature to conserve water reserves.

The harsh desert landscape is often obscured by violent sand storms

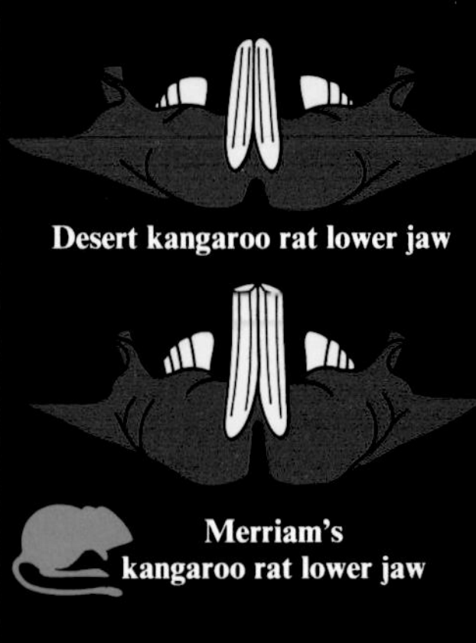

WATER REGULATION

Desert plants may be permanent residents or ephemeral opportunists. The animal kingdom in the desert displays a similar dichotomy.

Death Valley is famous for its desert bloom in the spring, although desert flowers tend to take advantage of good rains during any time of the year. Among animals the spadefoot toad (16D), common in the deserts of Sonora and Chihuahua, is capable of sealing itself up in a hardened mud-ball for years at a time, until summer showers finally release it from its self-entombment. The toad is also remarkable in always producing two sets of tadpoles, one small and numerous, the other larger and less numerous. The first set serves mainly as food for the second, some of whom thereafter survive to reproduce – an ingenious and efficient, if rather gruesome, adaptation.

Drought-resistant plants (xerophytes) display a variety of morphological adaptations. Succulents, such as the organ-pipe and saguaro cacti (18A), have thick, spongy tissues in which water can be stored, and leaves that have hard and waxy outer sides to prevent evaporation. Some xerophytes are instead covered in fine, hairlike excrescences that reduce air circulation on their surface, preventing loss of water and heat. Others may roll their leaves to reduce surface area exposure. They are often a light silvery-green to reflect the sunlight as much as possible.

16 17 18 19 20

A B C D

FAUNA AND FLORA

The life-forms of sandy-soiled deserts are subject to biological pressures different from those exerted by clayey hard pans (such as the Bonneville salt flats, in Utah) or of soils containing a high level of humus (as in the pinyon-juniper ranges of the Great Basin, Nevada). In every case it is the dominant vegetation that determines the nature of associated animal species. In shrubby deserts, for example, animals such as the black-tailed jackrabbit can find shelter under bushes, whereas in barren deserts only burrowing or crevice-inhabiting life-forms such as scorpions or fence lizards may be able to survive permanently.

Bushes help to create rich organic soils by filtering particles from the passing wind. Where competition for water is fiercest, however, some plants may exhibit allelopathy, in that they produce toxic substances in their leaves or other tissues which, when dropped, sterilize the soil where they fall, so inhibiting the growth of other seedlings. Such an emphasis on the production of toxins is often connected with defenses against grazing animals – defenses that are important in the austere desert conditions where food is at a premium. A prime example of allelopathy is provided by the sagebrush (6D), the most common bush of all in the Great Basin of Nevada, which not only sterilizes the surrounding soil but tastes disgusting to any animal unwise enough to try it.

RODENTS

The tendency of species to coexist by partitioning their available resources is especially important in an environment as forbidding as the desert. In order to survive, different species eat different kinds and sizes of food or feed at different times and in different areas. The Great Basin of Nevada, the Sonora desert, and the Mojave desert all contain remarkably similar ranges of seed-eating rodents, considered from the point of view of size and behavior, but the individual species are very different. It is very rare to find two different species the same size in the same area. The differences in body size mean that all of the species in an area can coexist by eating different-sized seeds. Merriam's kangaroo rat (18D) is widespread partly because of its ability to strip off and eat the outer, salty layer of the desert saltbrush (*Atriplex confertifolia*), without touching the plant's other, poisonous tissues. The chisel-like incisors that allow this are unique among its genus.

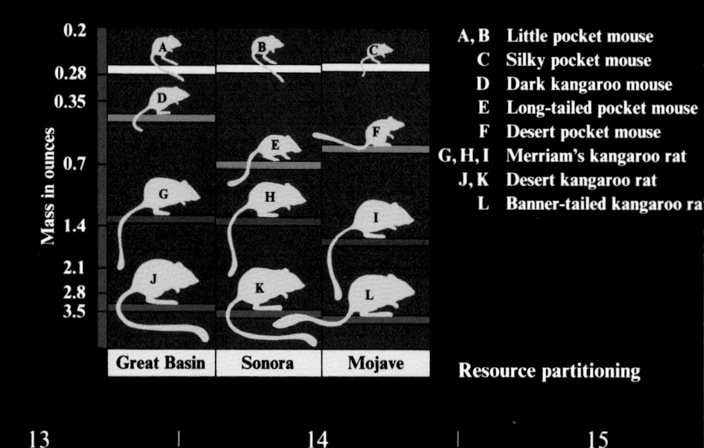

Resource partitioning

11 12 13 14 15

11 13 14 15

WILD HORSES

Wild horses (7C) roam freely in North America's western deserts, where they are often regarded as escapees and feral misfits. Yet the horse actually originated in North America, though horses and asses briefly died out there (about 7,000 years ago in the case of the horses, and about 12,000 years ago in the case of the asses). Native plants and animals are thus well integrated with the horses, which are merely refilling a niche only temporarily vacated.

The wild horses transport and scatter plant seeds in manure, and paw through the ice in winter, making forage or water available to other species such as deer (6C): temperatures of –4ºF are not uncommon in the winters of the North American deserts.

In summer, the mustangs paw through layers of thin shale and gravel to expose sources of water. During the dry periods of the year, mustangs depend on watering at least once a day, rendering them vulnerable to hostile ranchers whose waterholes and wells they may raid. During the cooler periods of the summer, the horses are able to go three or four days at a stretch without drinking at all.

The comparatively large size of the wild horse – like that of the bactrian camel and the burro – is itself an advantage in a desert environment, quite apart from making it less vulnerable to attack from coyotes and puma (1C). The reduced surface-to-volume ratio leads to a decrease in exposure to extremes of both heat and cold. In addition, wild horses have the habit on particularly hot or cold days of standing close together, side by side, usually in pairs, so that on hot days they shade each other and on cold days they radiate heat to each other. During fierce winter blizzards or on torrid summer days, the horses may shelter instead under juniper trees and pinyon pines (3C).

DESERT DYNAMICS

The key to survival

Most of the factors that disrupt other ecosystems contribute to the spread of deserts:

- Overgrazing by domestic livestock, often encouraged by misguided bore-hole irrigation schemes
- Soil compaction by animals' hooves, leading to water run-off and erosion
- Damming schemes which cause salinization
- Global warming
- Global wind patterns bringing descending, dry air to a region
- Rain shadows caused by mountains
- Violent winds carrying off fertile soil

Forces for change

- Agro-forestry projects and intensive water conservation, such as the drip-mulching in Nevada

Weather in the Sahara is depressingly predictable – hot sun, dry winds and no rain. But it has not always been like this. There are plenty of signs that the Sahara once experienced some pleasantly rainy periods, called pluvials. At the same time, other places in the world were suffering the bitter cold of the Ice Ages. During this time, substantial rivers flowed where now there are only rocks and sand dunes, and the Sahara's landscape, so dry and desiccated today, was shaped as much by water as by the wind. Even in Roman times, the Sahara was a fertile land of grass and trees, where grain was cultivated for the emperors.

The Sahara is not the only desert to have enjoyed wet periods. Indeed, there are signs that water once washed across most of the world's major deserts, helping shape the great sloping bajadas and giant gorges that are so much a feature of many desert landscapes. Equally, however, deserts have been through periods when they were even drier than they are now – times when the desert margins expanded far into areas that are now grassland or even forest. Beneath the grass and shrubs of the Kalahari bush in southern Africa, for example, are ancient sand dunes as big as anything in the Namib desert today.

In fact, the desert margins are constantly shifting backward and forward, not just over thousands of years, but from century to century, or even from year to year. A wet year encourages the hardier plants to creep into the desert, fringing it with green. A dry year withers such bold colonizers and expands desert conditions into areas that were once fairly moist. There is plenty of evidence that the world's deserts go through cycles of expansion and contraction in line with climatic fluctuations.

Desertification

In recent years, however, there has been increasing concern about the expansion of the Sahara desert right along its southern margin, from the Atlantic Ocean to the Indian Ocean, a region called the Sahel. Here, climatic fluctuations have combined with overpopulation by humans and livestock to create new deserts, a process called desertification because it now seems to be irreversible, even though there are a few instances of regeneration.

This process is going on all over the world – in India, China, Argentina, Chile, Mexico, and the Southwestern United States, and the formation of deserts has doubled in the last 100 years. Indeed, according to some authorities, an average of 40 sq miles of the Earth's surface turns to desert everyday. But the human implications of desertification in the Sahel, in particular, are immense. Millions of families live in these marginal lands, and when the land becomes too dry to farm, it can bring starvation on an enormous scale to countries like the Sudan and Ethiopia.

The Sahel disaster

In the past, people have been able to move away from these precarious areas whenever the climate became drier, then return when the climate improved again. But increasing economic and population pressure have made it much harder for the people of the Sahel to move away in drought years, and so they have been locked into an inescapable cycle of disaster and recovery.

During moister years, when rainfall is above average, nomads and semi-nomads used to drive cattle, sheep, and goats by the million into the dry grasslands that fringe the deserts. In places where the ground retained moisture from the wet season – on inland deltas, for example, and in the courses of some old rivers – people might even settle down in villages and start to grow crops on a subsistence basis.

In the moist years of the 1950s and early 1960s, both herding and farming

Desert dunes
The giant dunes that are found in Africa's Sahara and Namib deserts are virtually bereft of life, as they are too unstable for even drought-resistant plants to take hold.

moved northward into the Sahara. But when drought set in, in the 1970s and 1980s, overpopulation and economic pressure made it difficult to escape. Nomads were unable to take their herds south again – partly because of new international frontiers and partly because the land was already occupied by farmers growing subsistence crops. The subsistence farmers could not move either, as the best land was occupied by farmers growing crops for export.

Unable to proceed, the nomads' herds quickly begin to exhaust their normal grazing and turn to plant species they usually ignore, until these too are gone. Wild animals, likewise prevented from migration, have joined in the process. Farmers, meanwhile, are forced to carry on cultivating marginal soil until it too becomes impoverished. Overpopulation has forced people to remove more woody plants for fuel with which to cook.

Overcultivated and overexposed, the land is subject to wind and water erosion and begins a downward spiral to desertification. Once the soil deteriorates so far, it cannot recover even when the climate shows improvement. Moreover, desertification can actually make the climate drier. When the land is laid bare, its "albedo" – that is, its tendency to reflect sunlight – increases significantly. The extra warmth only serves to dry out the air even more, and so accelerate the desiccation of the landscape.

Although humans are playing an all too obvious part in desertification, the reasons why there have been so many drought years is unclear – and so there

is some doubt whether the desert margins will ever again be relieved by a succession of moist years.

In Niger, millet is planted in an attempt to stem the advance of the desert. In Algeria, the desert margins are being planted with specially watered trees, and it has been suggested that the whole desert might be ringed with trees.

Drought years

Some climatologists link the drought in the Sahel with the seasonal shifts in the Intertropical Convergence Zone or ITCZ. The ITCZ is the broad zone lying roughly along the equator where tropical trade winds blowing from the northeast and southeast converge.

In January, it crosses Africa far south of the Sahara, along the coast of West Africa. The Sahel is engulfed in a huge mass of dry air that sits over all the Sahara, keeping it bone dry. During the spring, however, the ITCZ swings northward, drawing moist air from the Gulf of Guinea with it. The Sahel is right on the limits of the ITCZ's northward migration, and depends on the moist air it brings for its summer rains. If the ITCZ does not travel quite far enough, the Sahel is deprived of its rain, and drought is inevitable.

Other climatologists link the droughts in the Sahel and the deserts of South America with the coming of El Niño, the cold ocean current that develops in the South Pacific around Christmas every decade or so. No-one knows quite what the link is, but it seems that the cold ocean could reduce the amount of moisture evaporating into the air and so reduce rainfall all around the world at the same latitude. The world's deserts are likely to be especially susceptible to global warming, if it should occur, and some predict that they may expand significantly over the next century.

In many places, people have tried to reverse this trend by large-scale irrigation, and vast areas of marginal land are now on artificially-watered croplands. However, there are problems. Moisture evaporates rapidly in arid areas, drawing toxic salts to the surface and killing plants. Salinization has already ruined large irrigated areas of India, Syria, Iraq, central Asia, and the San Joaquin Valley in California. More worryingly, such schemes often rely on stocks of water held underground since the pluvials. Now these too have shown signs of severe depletion.

Arid relief

Sometimes it rains even in the driest desert. But in between storms, the desert surface dries so hard that water cannot sink in. So the rain, when it does come, can actually bring widespread floods (top). After the rain, the desert blooms briefly with plants that have stayed dormant through the drought (above).

AROMATIC PLANTS

Many aromatic plants (10D) flourish in the maquis. Indeed, after rainfall, the whole landscape is suffused with the powerful fragrance of herbs as the hot sun evaporates water from the foliage, along with traces of the aromatic volatile oils produced by the plants. An aromatic plant (such as lavender, marjoram, rosemary, sage, or thyme) has many tiny oil glands dotted all over the surfaces of the leaves, stems, sepals, and petals. The dual purpose of the glands is to provide an oily sunscreen against scorching solar radiation and, perhaps more importantly, to deter grazing animals, for most animals avoid plants with a sharp flavor if there is a more palatable alternative nearby. In summer, though, when much of the tender vegetation has toughened or shriveled up, animals may have to eat aromatic plants or face starvation.

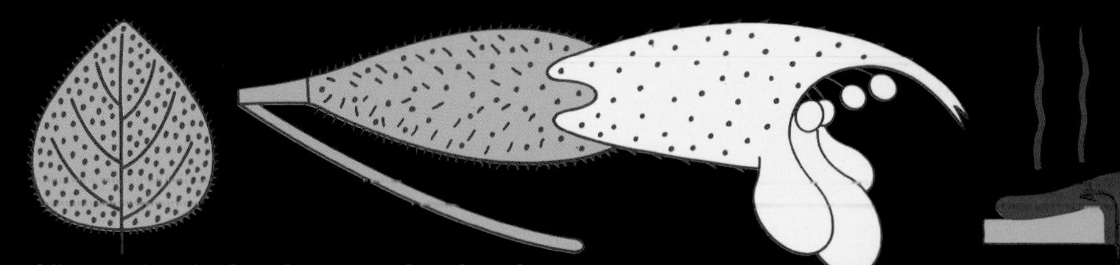

Oil producing glands on leaves, sepals and petals

Aromatic vapors

6 | 8 | 9 | 10

6 | 7 | 8 | 9 | 10

nourishment from the tough cellulose in the tissues of the plants consumed. Goats especially are ingenious feeders, even climbing trees in order to browse.

Lagomorphs such as the rabbits and brown hares employ a different method of finding sufficient nourishment in the cellulose. Food travels the length of the alimentary canal, is excreted only partly digested, and is swallowed again for a second passage down the alimentary canal.

SURVIVING THE SUMMER

In the Mediterranean region plants must survive a near-total drought that may extend from May right through until September. One method of obtaining water is a taproot, stout and unbranched for much of its length, penetrating deep into the ground, and becoming branched and fibrous only once a permanent supply of groundwater has been reached.

To conserve water, some species have developed subterranean storage organs such as bulbs, corms, and tubers. Such plants often become dormant during the summer months, when the underground organ stores water and carbohydrates which enable the plant to remain alive at a much reduced metabolic rate. Many plants are annuals, completing their entire life-cycle within a single growing season. In the summer the seeds simply lie around in cracks in the dry earth, requiring no water, waiting for autumn.

The leaves of sclerophylls have a thick, waxy epidermal surface which impedes the passage of water. To lose less water, leaves in other plants may be so small that they cease to resemble leaves at all – like the scale-leaves of some juniper species. By rolling up, grass leaves stave off water loss because their stomata (breathing pores) are on the inner surface and are thereby sheltered from sun and wind. A few plants go to extremes and actually shed their leaves during the hot months. Such summer-deciduous species include the tree spurge.

The landscape typical of the Mediterranean basin is dominated by maquis scrub – few trees, but an abundance of shrubs and smaller plants. The area is rich in mammalian life, from small rodents to herds of sheep and goats; there are also reptiles, an ever-changing flux of resident and migratory birds, and a great variety of insects. The wilder, more fearsome forces of nature are not well represented here, but the maquis ecosystem is nonetheless beautiful and complex in the multiplicity of interactions between its species.

INSECT LIFE

The ecosystem benefits from an abundance of insects fulfilling many necessary functions, in particular either as pollinators of plants or as part of the food chain. Two that are perhaps characteristic of the maquis are the scarce swallowtail butterfly (1D) and the dung beetle.

The scarce swallowtail butterfly belies its name. It is in fact widespread throughout the region. Females lay eggs on fruit trees so that on hatching the larvae can eat the succulent leaves. Another butterfly, the southern festoon, only lays its eggs on birthworts (*Aristolochia* spp.).

The dung beetle plays an absolutely vital role in the regeneration of animal waste. From a deposit of dung, the beetle fashions a ball of manageable size and rolls it along the ground, pushing with its hind legs. At its burrow, the beetle buries the ball of dung in an underground chamber, and lays an egg in it. The larva that subsequently hatches feeds on the dung.

GRAZING ANIMALS

The maquis ecosystem is subject to an often heavy grazing pressure which prevents the vegetation from developing into woodland. By far the most voracious grazers are goats (7A) and sheep (4B), but also responsible are rabbits, brown hares, and, to a lesser degree, tortoises (15D).

Sheep and goats are ruminants – animals in which the stomach is divided into four compartments: the first two compartments are for storing hastily swallowed food mixed with saliva before it is regurgitated back into the mouth for proper chewing and reswallowing for digestion in the third and fourth compartments. The diastema, the gap between the incisors and molars, allows the thorough mixing of food and saliva. Because the chewed food is already softened, the jaw muscles of ruminants are weaker than those of, say, horses. The comparatively powerful cheek muscles move the jaw up and down, while other facial muscles make lateral, grinding movements. This second chewing to grind the regurgitated food down into a pulp is generally undertaken while the animal is resting (and in cattle – also ruminants – is described as chewing the cud); only the fourth and final compartment has gastric glands that secrete digestive juices. This lengthy and complex process is necessary in order to break down and get sufficient

Sheep's head

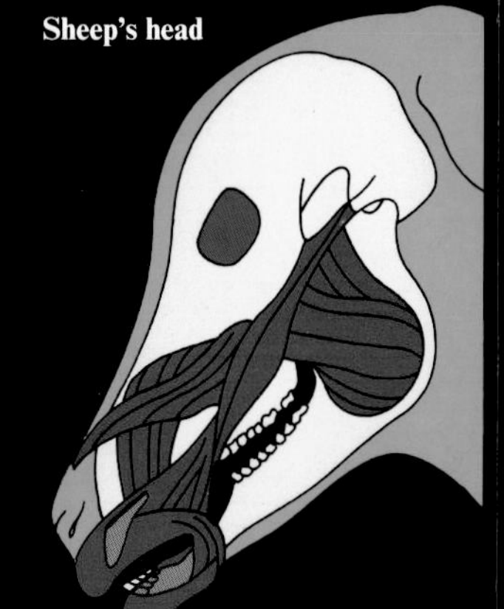

SCRUBLANDS

In the lands around the Mediterranean Sea, summer skies are almost always blue and clear, and hot sunshine beats relentlessly upon the landscape, parching the ground and stunting plant growth. Trees are rarely large in these arid conditions, and vegetation is often reduced to a thorny scrub of shrubs and aromatic herbs, especially where the land has been heavily grazed. But Mediterranean lands are only hot and dry in summer; nearby seas mean that winters are cool and moist, bringing fresh growth and allowing a rich variety of plants and animals to flourish.

Scrublands like this occur on every continent in the world – wherever hot summers combine with cool, moist winters to create a Mediterranean-type climate. They go by different names in different places – mallee in Australia, chaparral in California, fynbos in South Africa, and mattoral in Chile. But in each region, scrubland plants and animals alike are specially adapted to sustain the ecosystem through the ravaging drought of high summer.

The winter rains that often disappoint late tourists are vital to the survival of Mediterranean scrubland life. Without the cool refreshment of winter, the almost rainless summer months would reduce the landscape almost to a desert. In cool temperate regions, plants tend to grow in summer and lie dormant in winter; in Mediterranean scrubland, the reverse can be true. The year's first new shoots may emerge not in spring but with the first rains of autumn. Ahead lie all the mild, moist days of winter in which they can grow strong and sturdy before flowering and fruiting in the warmth of spring.

Summer, by contrast, is a time of survival. The hottest months average over 68°F, and temperatures can soar to well over 100°F, baking the ground hard. Some annual plants die off completely and get through the summer as seeds. Some, such as fritillaries, grape hyacinths, and cyclamens, make it through the summer drought only by dying back to an underground storage organ such as tubers or bulbs.

Water-saving

Many of the shrubs are deciduous, shedding leaves to save water just as cool-temperate deciduous trees do in autumn to avoid winter frosts. Others, such as the proteas of South Africa and the banksias and hakeas of Australia, are evergreens, called sclerophylls (Greek for "hard leaves") because they have small tough, leathery leaves well able to withstand the desiccating heat of the summer sun.

Some shrubs have roots so deep that they can tap into water stored far underground and can carry on growing throughout the year. This may give them the competitive edge over trees and grasses but lays them open to the attentions of grazing animals and insects anxious for moist nourishment in the summer drought. To counteract this, some species are armed with tough leaves and thorns; others, like chamise, resort to chemical warfare, exuding strong-smelling juices to deter would-be diners.

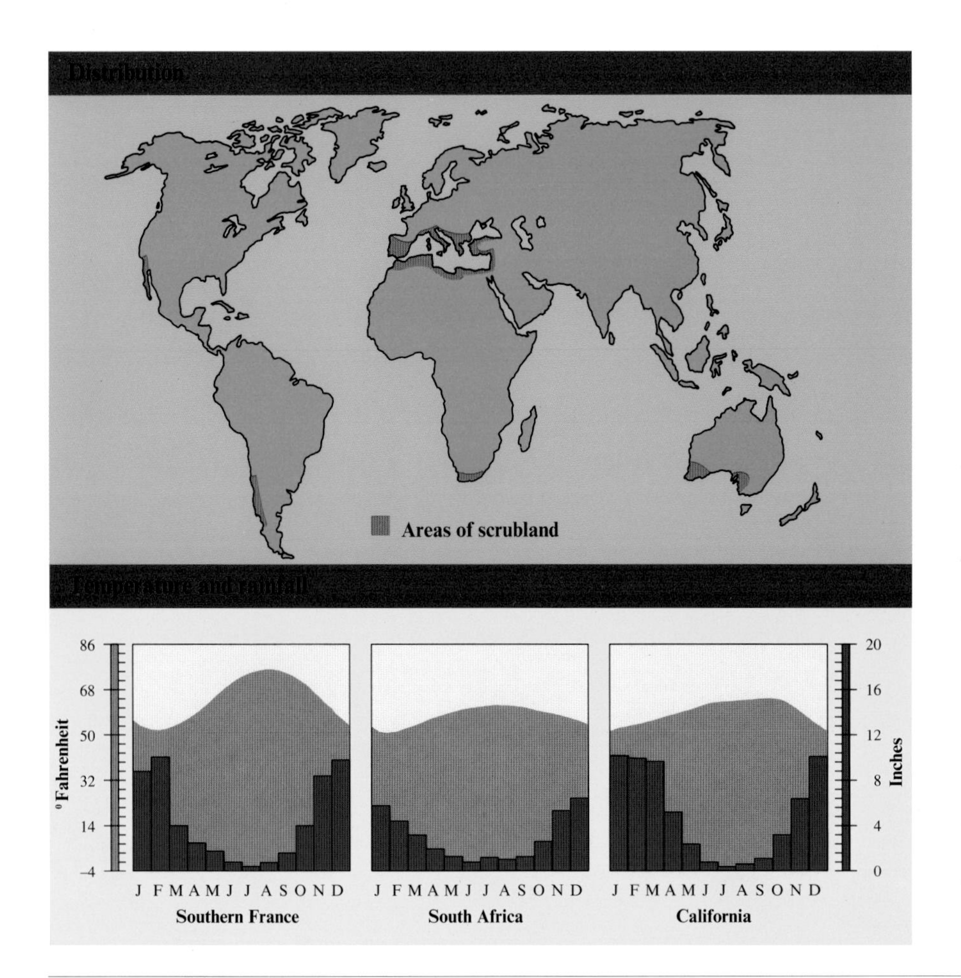

Explosive juices

The aromatic juices in shrubs also help to inhibit the growth of herbs, which would otherwise provide strong competition. But they are highly flammable and, during the dry season, shrubs will nearly explode when set alight. Every now and then raging fires rip through the scrub, clearing away old shrubs, burning off these toxic chemicals and releasing nutrients into the soil and atmosphere.

After the fire, herbs and grasses shoot up, providing good food for deer, sheep, and cattle. Fire is thus an integral part of the scrubland ecosystem, continually clearing the way for new growth. Where humans have tried to prevent fires, they can disrupt the natural balance, allowing shrubs alone to flourish.

In the Californian chaparral, one of the dominant shrubs is sagebrush, which exudes aromatic juices called terpenoids. Because these are so volatile, the sagebrush is completely destroyed by fire every 25 years or so. The fire takes with it any shrubs that might otherwise compete with the sagebrush. New herbs spring up after the fire, but sagebrush is beginning to invade again within three or

The protea of the South African fynbos: a perfectly adapted arid-zone plant

four years. Ten years after the fire, large patches of sagebrush have developed and the terpenoids they exude have killed off all the herbs, leaving the ground bare. Herbs only survive in the patches between the brush.

Scrubland trees

In moister areas, there are often ragged woodlands of pines, cypresses, or even oaks such as the summer deciduous white oak and the evergreen holm oak. Natural oak forests were once widespread, but in the Mediterranean and even California they have now been largely cleared for farming, and olives and vines grow where once there were oak trees.

Where conditions are especially dry or windy, or soils are stony, tree growth is stunted and scrub is the natural climax vegetation. Sometimes, especially on coasts and islands, not even shrubs grow, creating an open, steppelike landscape of grasses, small herbs, and bulbous and tuberous plants.

Often, though scrubland is not natural but the result of forest clearance. In the Mediterranean, the long degradation of pine forests has left a kind of scrubland called garigue, dominated by aromatic evergreen shrublets, while cork oak woods have degraded into maquis, dominated by dense evergreen sclerophyllous shrubs. The ubiquitous goat ensures that the trees rarely grow back.

Scrubland fauna

Even degraded land like this, though, supports a surprisingly rich array of creatures, especially small mammals (such as seed-eating rodents) and reptiles, such as lizards and snakes, which may be seen basking in the sun. There are also many insects, notably the cicada, which chirps noisily throughout the summer. Very occasionally in Europe larger mammals such as the roe deer, lynx, wild cat, and wolf may still be seen. In Australia, there may be kangaroos, in South Africa hyraxes and duikers, and in North America jack rabbits and pumas. There were once even wild lions and elephants in the Mediterranean and North America, but these have long since been exterminated by humans.

Not all creatures are visible all year round. Some only move into the scrubland in winter. The Californian chaparral, for example, is teeming with birds and other vertebrates during the wet season. But many of these creatures depart for greener pastures during the long hot summer. Little groups of collared peccary, for example, retreat to the cool of the forest every spring. Ground squirrels and kangaroo rats, on the other hand, retire to their cool burrows beneath the ground. Here they store the seeds they have collected during the spring, and these seeds may help them conserve moisture by absorbing water vapor expired by these rodents underground.

Some even very small Mediterranean islands once had their own unique fauna of mammals, typically dwarf races of animals found on nearby continents, such as dwarf elephant, hippopotamus and deer, all of which were presumably wiped out by early human settlers, like the lions. Yet, interestingly, before the arrival of the humans, these creatures had no real predators. On the island of Crete, for example, no carnivore fiercer than a badger is known to have existed – and in the past herbivores must have lived there in considerable numbers, limited only by the availability of food.

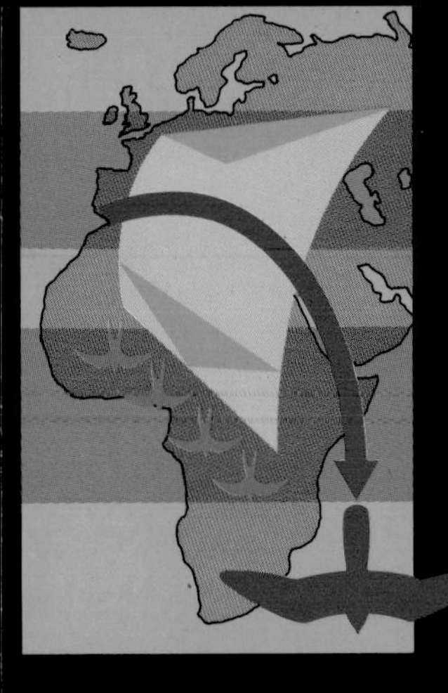

Eleonora's falcon and its migratory prey

MIGRATORY BIRDS

Large numbers of birds visit southern Europe every summer, adding greatly to the diversity of the maquis scrub as they pass through the Mediterranean from their wintering grounds in Africa. Some species fly straight across the area, stopping perhaps only for a few hours on islands such as Sicily and Crete, where many are shot for food. However, a number of summer visitors remain in the Mediterranean area throughout the season, to nest and to rear their young before returning south in the autumn.

The woodchat shrike (13A), a small but striking bird, is one of these breeding summer visitors. From its prominent lookout perch amid scattered bushes it swoops to catch larger insects and invertebrates which it impales on sharp thorns – a practice common to all shrikes, and one that has earned them their description as "butcher birds": the thorny twig is the bird's "larder."

Eleonora's falcon breeds only on the ledges of inaccessible coastal cliffs on some Mediterranean islands, notably the Balearic Islands, Corsica, Sardinia, Crete, and Cyprus. The hatching of its young coincides with the migration of huge numbers of smaller birds south across the Mediterranean in late summer and early autumn. These birds provide an abundance of food for the young falcons.

16 | 17 | 19 | 20

A | B | C | D

17 | 18

REPTILES OF THE MAQUIS

Because they are unable to regulate their body temperatures internally, the hot, sunny expanses of maquis scrub provide ideal habitats for several types of reptile. There are patches of bare ground and outcrops of rock on which to bask, and dense prickly bushes under which to shelter from the cold at night, escape predators, rear young, and hibernate. There is also plenty of food available in the form of plants, insects, small mammals, and even other smaller reptiles.

Young green lizards (2C) are actually a nondescript brown, becoming bright green only when mature. Their quick, scuttling reactions help them stay out of the clutches of birds and snakes.

Hermann's tortoise (15D) may live to a great age, possibly more than 100 years. It hibernates in winter, either buried underground in loose soil or sand, or in a sheltered hollow beneath a dense shrub. The prime imperative for a male tortoise on emerging from hibernation in spring is to locate and impregnate a female. Only then does it begin to search for edible plants, or its preferred diet of small invertebrates, to replenish the reserves of fat depleted during the winter.

The Montpellier snake (19D) and the southern smooth snake are two species characteristic to Mediterranean scrub communities. Although venomous, the Montpellier snake has no need to use its poison on smaller prey, which it swallows very quickly. The smooth snake is a true constrictor but rarely kills by constriction, instead holding the prey steady in its coils in order to swallow it. Both feed on lizards and small mammals, and the occasional bird.

The chameleon (18A) is slow-moving, but can change its color to blend in with the vegetation, and so creep up unseen on insects. Once a victim has been located, the chameleon flicks out a long, sticky tongue to catch its prey.

SHRUBS AND TREES
Shrub and tree species characteristic of the maquis include brooms, cistus (rock roses), dwarf fan-palm (15D), French lavender, and heathers, all of which flower brightly. In spring, whole areas are stained yellow with broom and Jerusalem sage, pink and white with cistus, and white with tree heather. Other species flower in autumn, like the strawberry tree, which has small, creamy bell-shaped flowers and red, strawberry-like fruits that take 12 months to ripen.

The maquis ecosystem also includes a number of potential tree species, such as kermes oak (19C), holm oak, wild olive, and carob (locust bean). A few of these may succeed in growing into fully developed trees rising out of the surrounding scrub, but more often they are bitten down by the grazing animals, sometimes into gnarled living structures. Amazingly, the kermes oak is able to flower and to fruit (and therefore reproduce) even in this bitten-down state.

RESIDENT RAPTORS
The griffon vulture (10A) is a bird of hill and mountain countryside that contains cliffs. Large, vertical rock faces provide nesting sites and, very importantly, thermal upcurrents in the air which enable this huge bird to soar high into the sky, where it remains for long periods. Maquis scrub communities often surround cliffs and mountains in the Mediterranean basin, and sheep and goat fatalities are valuable sources of nourishment to the vulture. One of the few situations in which griffon vultures are seen *en masse* is in the clamorous throng that descends from the skies to pick at a carcass.

The imperial eagle – another bird of prey – is a relatively rare species in the Mediterranean basin. A ponderously heavy-looking bird, it inhabits open land, hunts over fields, plains, and marshes, and nests in isolated trees. Although imperial eagles from farther north migrate annually, those in the maquis are permanent residents.

THE BEE ORCHID
The Mediterranean basin is the center of distribution of a very specialized group of orchids that have evolved flowers intended to resemble female insects. They include the bee orchid (16C), the spider orchids, the wasp orchid, and the fly orchid (all *Ophrys* spp.).

Male insects may try to mate with the flower, when the orchid's sticky pollen sacs become attached to the insect's head. The unsatisfied insect then flies off, possibly straight to another orchid, to which it transfers the pollen sacs.

As is evident in the names, one specific insect tends to be the pollinator only of one specific type of orchid. Inspite of this specificity, all the orchids nonetheless seem to be in a state of active evolution in the Mediterranean basin. Each species has numerous local races or varietal types, each one identifiable by slightly differing features, including shape and color. Occasionally the difference is not in shape or color but in flowering season.

SCRUBLAND DYNAMICS

The key to success

- Grazing by sheep and goats and, to a lesser extent, by hares, rabbits, and tortoises
- Controlled cutting for brushwood
- Controlled burning to create new growth for livestock
- Constant low level of precipitation

Forces for change

- Uncontrolled fire, started by lightning or by human agency
- Violent storms that fell trees and uproot large shrubs
- Uncontrolled clearance for agriculture
- Construction of buildings and roads

The maquis scrubland ecosystem of the Mediterranean is not natural. If protected from grazing sheep and goats and, to a lesser extent, hares, rabbits, and even tortoises, it eventually regenerates into the Mediterranean evergreen forest that is the natural climax community for the region. Maquis areas often contain species, such as the kermes oak, as bitten-down shrubs. If grazing ceased, these shrubs would shoot up to become full-grown trees within a few decades. The seeds shed by those trees that do exist would also be able to grow to maturity unchecked.

Each animal has its own method of grazing. Sheep, for example, tend to pick out the young, juicy, tender shoots of dwarf shrubs and other plants close to the round. In contrast, goats are far more adventurous and eat a much wider range of vegetation, including plants with rough textures and even prickles. Goats are agile climbers too, and climb both cliffs and trees to reach edible foliage when there is no more easily accessible food. Large numbers of goats concentrated in one area of maquis can be highly destructive to the ecosystem, and may eventually reduce it to almost steppelike conditions. But they dislike strong flavors, and it is only as a last resort that they are prepared to eat resinous plants such as pine, juniper, and lentisk (the mastic tree) or sharply aromatic species such as savory, thyme, and French lavender. Species like these tend to dominate heavily goat-grazed areas because everything else has been consumed.

Potential of tree growth is also restricted to fairly shrubby proportions by the cutting of maquis shrubs by villagers, farmers, and shepherds to produce brushwood for making hedges to enclose fields, and for use as kindling to light fires. In addition, shepherds burn the maquis vegetation periodically to maintain sheep pasture. Their usually well-controlled fires prevent the maquis from becoming impenetrable and allow sunlight to reach the soil, so enabling tender new shoots of shrubs and other low plants to grow.

A further, much less obvious, factor supporting maquis communities is rainfall levels. Heavy rain may limit the effect of grazing, clearance and fires, and allow woodland to regenerate. Low rainfall, on the other hand, may reduce the maquis to an open steppe-like community with few shrubs, dominated instead by grasses, bulbs and herbs.

Natural disasters

The maquis ecosystem is surprisingly unaffected by most natural disasters, except for fire. Fire can rip through the dry maquis vegetation very quickly completely destroying all vegetation above ground level and leaving carbonized stumps of shrubs and trees. Yet recovery can be rapid. Plants that have subterranean rootstocks, bulbs, and tubers are not permanently affected and will regrow. Where seeds lie sufficiently deep beneath the surface not to have been singed, they will still germinate. And from the charred bases of some species of tree (such as kermes oak and olive, but not pine), shoots will sprout. Burnt maquis eventually, and more or less completely, regenerates.

Pine trees in particular are highly inflammable because of their resin content and once pines are established in any area, a fire is inevitable. Indeed, fire is an essential part of the biological regime of certain species of pine, like the Aleppo and Calabrian pines. These are invasive species, and colonize rapidly, partly because they are so unpalatable to grazing animals. Pine actually relies on fire for regeneration. New pines start not as shoots from the stumps but from seeds that germinate after exposure to the heat of fire. The land thus becomes locked into a recurrent pine growth/fire/growth cycle.

Unnatural disasters

In places. however, farmers have completely destroyed maquis by clearing even the shrubs for growing crops such as cereals, vines, olives, fruit trees, or salad crops. Large scale scrub clearance usually involves bulldozing, which disrupts the soil structure and permanently alters the local topography. Furthermore, modern agriculture in the Mediterranean basin frequently uses (or misuses) chemical fertilizers and herbicides which may damage the ecosystem even beyond the agricultural area.

The maquis is not destroyed permanently even by total clearance. Left alone for long enough, maquis regenerates on agricultural land. Areas cultivated in the past (identifiable sometimes by terracing that is still visible) may today be seen at various stages of regeneration – a few scattered bushes, dense maquis, and even climax (ecologically stable) woodland.

It is the construction of buildings and roads that brings about the most radical and most permanent changes of all. Not only is the land heavily bulldozed, but it is sealed – possibly for centuries to come – in concrete or tarmac. Yet a degree of regeneration may be possible even in these extreme circumstances. Large stone structures dating from ancient times, such as 2,500-year-old Greek amphitheaters and temples, are gradually colonized by plants after they are abandoned. The real extent that steel-reinforced concrete can ever be colonized by vegetation, and the length of time it takes to happen, remain to be seen.

Mediterranean cliff communities

Clambering up the deep gorges and steep cliffs characteristic of limestone areas in Mediterranean lands are plants known as *chasmophytes*, which are well adapted to life on cliffs with little soil, very little water, and high daytime temperatures. Few other plants can tolerate conditions like these, so the *chasmophytes* thrive here, and different species have evolved to exploit particular situations. But this extreme specialization means that *chasmophytes* are unable to compete in other habitats, and are found only in sites so inaccessible that they have no need of the spines and other weapons evolved by most Mediterranean plants to protect themselves from foraging animals.

The fynbos in South Africa with mature shrubs

Shrubs destroyed by fire

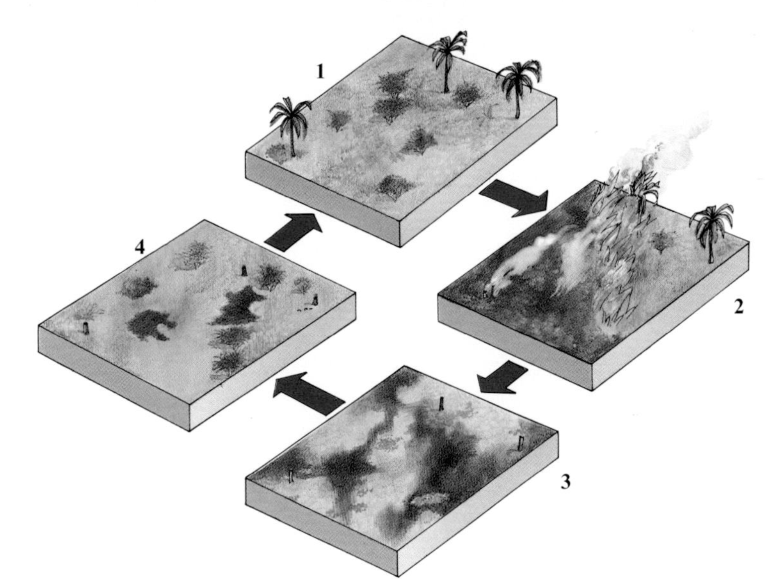

Herbs and flowers spring up after a fire

Fire in the maquis
Maquis scrub (1) varies in density from low, widely-spaced bushes to dense thickets of tall shrubs. It is prevented from reaching its natural evolutionary climax – evergreen forest – by grazing, cutting, and firing.

When fire breaks out (2), denser maquis burns more freely than open scrub. Resinous plants (such as pine) or shrubs that contain volatile aromatic oils (such as lavender) burn intensely. Fast-moving animals such as sheep, goats, hares, rabbits, lizards, and snakes, together with birds, escape the flames. Slow-moving creatures such as tortoises perish.

A few years later, some trees and shrubs sprout new shoots on charred stumps (3) from buds that were protected beneath the soil from any fire. Buried seeds, bulbs, and tubers also produce new growth, especially of herbs that are normally suppressed by the shrubs. Grazing animals have also returned.

Some 20 years on from the fire, the maquis has almost completely regenerated, and a few charred stumps are the only traces of the flames (4). The herbs that shot up briefly after the fire have gone and a few plants whose seeds, bulbs, or tubers were unable to grow after the fire may never have returned.

FISH HYDRODYNAMICS

The shape of a fish has a great influence both on its swimming pattern and on its way of life. Elongated fish, such as the moray eels (6E), swim by an undulation of the body and hide in the cracks and crevices of the reef.

The butterfly fish (8A), with its laterally compressed body, relies on its tail to provide the main source of forward thrust, and paired fins at the front to provide rapid deceleration when fully extended. The paired fins also make for delicacy and control in swimming movements, enabling the butterfly fish to maintain position in swirling water currents. Its slender shape allows it to dart rapidly between the branches of a coral colony at the approach of a large predator.

The parrotfish (11E), which gets its name from its strong, parrot-style beak, is adapted for feeding directly on coral. It bites off the branching tips and grinds them with platelike teeth at the back of its mouth. The soft parts are digested, and the skeleton is excreted in the form of fine sand. Many parrotfish prefer to eat algae scraped off the coral.

6 | 7 | 8 | | 10

2 0-20 feet deep

| 8 | 9 | 10

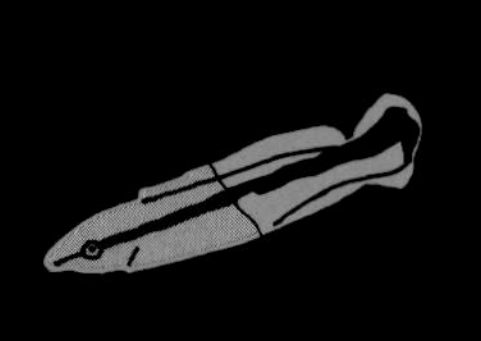

Scale-eating blenny: swimming profile

Cleaner wrasse: swimming profile

PARTNERS AND MIMICS

The cleaner wrasse, *Labroides dimidiatus* (14G), is a small, brightly colored fish resident among coral reef communities. It provides a service by removing parasites from other fish. Its striped blue coloration and its characteristic swimming pattern serve to bring it to the attention of larger fish, which may then wait in line for its offices. Even the mouths of large predatory fish such as moray eels are opened for the cleaner wrasse to remove particles from between their teeth.

The scale-eating blenny, *Aspidontus*, makes its livelihood by mimicking the cleaner wrasse. The blenny's color changes according to its mood, and when it is aggressive or hungry it not only has a very similar coloration to the wrasse, but also swims in the same undulating style. Unsuspecting fish approach and hold still in front of the blenny, mistaking it for a cleaner, whereupon the blenny lines itself up on a target area, darts in – usually from behind – and bites out a piece of the larger fish's fin.

The remarkable resemblance between the cleaner wrasse and the scale-eating blenny is just one of a number of examples of mimicry to be found in the complex community of a coral reef.

Symbiotic and commensal (literally "feeding at the same table") relationships are extremely common in coral reef communities. One of the best known is the relationship between the anemone fish and its anemone (4C) "host." The fish lives among the anemone's tentacles and nematocysts (stinging cells used to paralyze prey) without suffering any apparent harm. It is protected from predators who avoid the anemone, although the anemone derives little or no benefit from the presence of the fish. Different species of anemone fish live in different species of sea anemone, but all are brightly colored, with bold black bands.

Ecosystem Profile *Indo-Pacific Reefs*

Coral reefs are at their most diverse in the western Pacific, at the junction between the Indian and Pacific Oceans, where as many as 400 different species of coral are to be found. Not only is this a center of coral variety, but it is also a treasure-house of organisms that make reefs their home – the fish, mollusks, and echinoderms. The diversity of all these life-forms diminishes in proportion to distance away from the area, eastward across the Pacific or westward across the Indian Ocean.

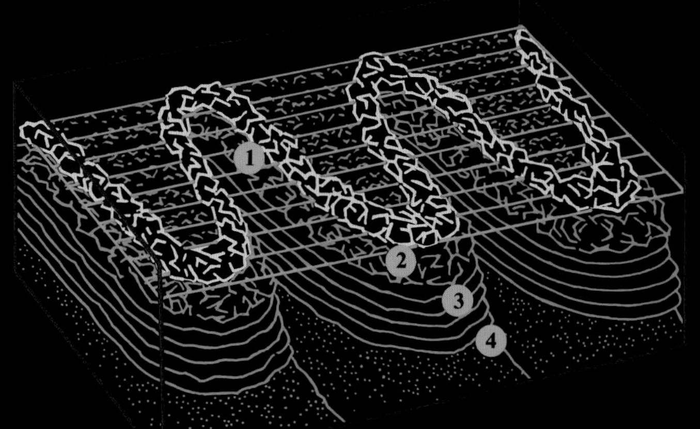

Fringing reef: showing picture sites

SURGE CHANNELS

The upper face of a coral reef, below the crest, at which wave action is strongest, is covered with platelike and heavily branched coral colonies. Where there are powerful ocean swells and currents, the reef's outer edge is not normally smooth but broken by narrow channels that cut deep into the reef itself. Inside these surge channels wave energy tends to be much less than on the exposed reef face, and the channels accordingly represent comparative shelter for communities of organisms quite different from those that reside on the reef face. The restricted width of the channels means that light penetrates only to a specific depth, so many of the deeper-water organisms tend to live closer to the surface in the surge channels. Sea fans (13E) and sea whips, for example, may lurk on the underside of overhangs quite close to the surface, while on the sides of the channels filter-feeding animals such as feather stars (9E) sift materials from the water as it flows past.

The symbiotic relationship between corals and their zooxanthellae (algae) is extremely important for the growth of the coral reef colonies. Skeleton formation in the reef-building corals is achieved at a more rapid rate, thanks to the presence of the zooxanthellae, than in solitary species which lack these symbiotic algae. The unicellular algae reside not only inside the coral tissue but, in some species, also on all the exposed surfaces of a reef, where they may become food for grazing mollusks which scrape them off the surface. Giant clams (1D) also harbor such symbiotic algae in their tissue. Multicellular filamentous algae may grow to form a fine turf, which may in turn be nibbled and eaten by grazing fish.

The algae associated with the coral help to consolidate the surface of a reef: their encrusting growth cements the dead coral skeletons together, providing a stable surface for new coral growth. In contrast, there are destructive phases in a reef's development. Sponges, worms, and other organisms may bore through the coral skeleton, weakening it so that bits may break off during storms, sinking to the base of the reef structure to form a scree slope, or be cast up onto the reef surface where they contribute to its overall upward growth.

CORALS

Azure blue tropical waters and feathery fronds of coral of all colors and shapes make coral reefs perhaps the most beautiful of all natural habitats. By day a myriad rainbow-colored fish dart through the coral, shimmering in the sunshine. By night, all kinds of bottom-living creatures – mollusks, sea urchins, brittle stars – emerge, revealing yet another part of the astonishing diversity of this marine ecosystem.

Yet what makes coral reefs truly remarkable, perhaps, is the fact that corals are not plants, but colonies of animals, living and growing together in a unique way. As individual corals die, their skeletons remain, and new generations grow on these to build up huge, intricate structures. The colonies' growth, though, depends on a close symbiotic relationship with small algal cells living in their tissues. Every coral colony is thus a combination of both plants and animals, and may be more accurately described as a "coral-algal" reef.

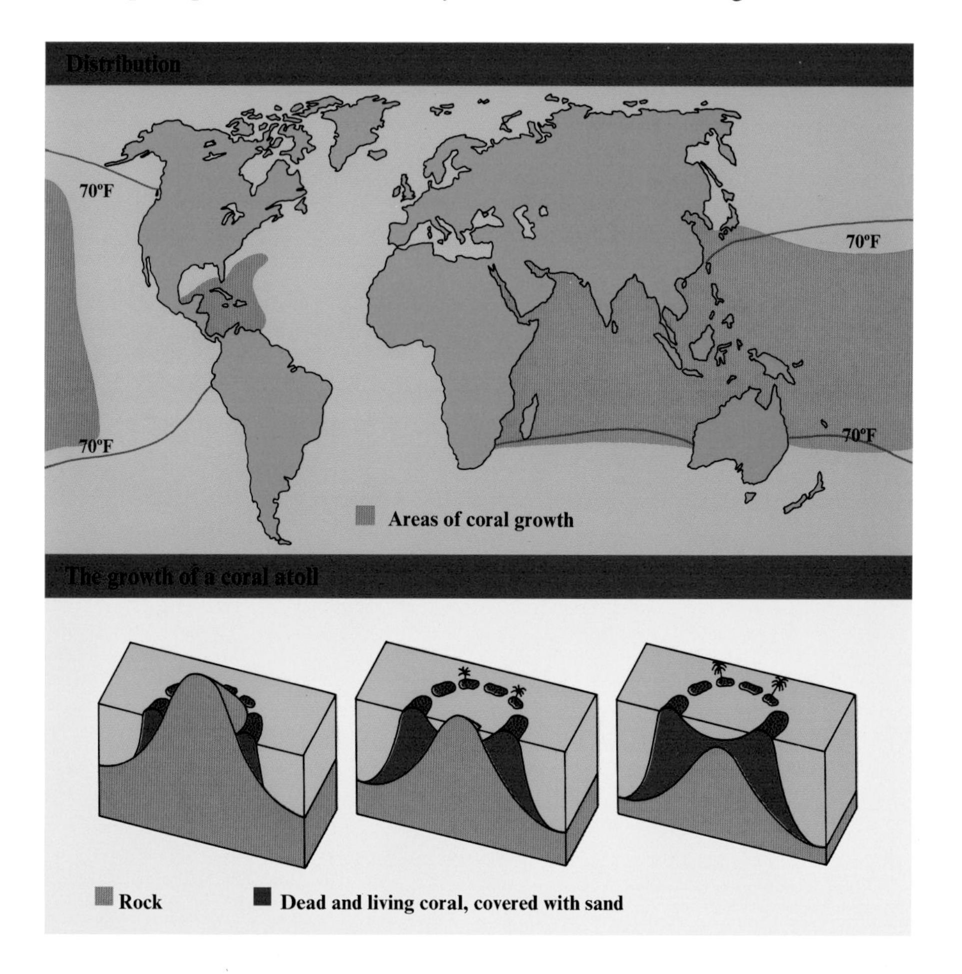

Corals are very particular creatures. They thrive only in warm water – water where the temperature during the coldest month averages at least 75°F. So reefs tend to occur only in tropical areas. A few solitary corals grow in colder waters at higher latitudes, but proper reefs are found only in warm water. If the water temperature ever drops too far, the colonies will die, as they did in 1968 when the temperatures in the Arabian Gulf plunged to 50°F, wiping out most of the inshore reefs. They will also die, however, if the water gets too hot, as it did in the Indian Ocean in 1987, killing off huge quantities of coral around the Maldive Islands.

Because their symbiotic algae need sunlight for photosynthesis, reefs occur only in shallow, clear water. Corals do not grow much deeper than 65 to 100 feet, and usually grow most vigorously just below the surface, where the water is well oxygenated. But they will not grow higher than the lowest tide level of the sea, because they need to be permanently immersed in water.

Types of coral reef

There are four major classes of coral reef structure: fringing reefs, barrier reefs, atolls, and platform reefs. As their name suggests, fringing reefs fringe the shoreline – typically around a continent or a volcanic island. Because corals are intolerant of fresh water, they tend not to occur in river estuaries, and ships can often find passages through the reef opposite river mouths. A fringing reef protects the shoreline from storms and tempestuous waves by absorbing the wave energy in its complex structure, so also minimizing coastal erosion.

A barrier reef occurs farther from the coast than a fringing reef and is separated from it by a lagoon. It was Charles Darwin who saw that if the land continued to sink, or the sea level to rise, a fringing reef might ultimately form a barrier reef, for coral growth is most vigorous at the reef crest (the seaward edge of a living reef community). Here, the corals grow upward and outward much faster than in more sheltered landward areas, where the oxygen content of the water is lower, the water temperature higher, and the availability of plankton food is less. As the land sinks, the outer edge of a fringing reef continues to grow, keeping pace with the rising sea, until a deep channel or lagoon eventually separates the land and the outer reef edge – which is now a barrier reef.

Atolls are ring-shaped structures in which a living reef surrounds a central lagoon. They are characteristic of oceanic areas away from continental landmasses. Charles Darwin, during the voyage of *The Beagle*, suggested that atolls formed as a reef that was growing on the top of a submerged volcano. His theory was proved correct a hundred years later, when deep drilling at Enewetak atoll in the Marshall Islands showed that the living reef was growing on the consolidated limestone formed of the skeletons of coral colonies that themselves had grown on top of a volcanic seamount. The thickness of the underlying coral limestone at Enewetak has been measured at 4,500 feet.

Platform (or patch) reefs are structures that are circular in area. They may constitute part of a barrier reef system, or may be located in oceanic waters where, like atolls, they are found on top of seamounts that are too small for an inner lagoon to form.

Clusters of platform reefs and atolls can be found together in the Indian Ocean

The form of coral colonies

The shape of different coral colonies varies widely even in different areas within a single reef. Crossing a fringing reef seaward from the land, an observer can quite clearly pick out the zones of different coral growth. In areas of high wave energy, the colonies tend to be low and compact, typically formed of brain corals. Enjoying relative shelter behind the reef crest are the colonies of finely branched species such as *Stylophora*.

The reef flat, or back-reef, behind the reef crest is itself highly variable, extending only a few yards behind the crest or stretching for several hundred yards, reaching the shore on fringing reefs and descending into the lagoon of barrier reefs and atolls. The more extensive reef flats show a pattern of zones every bit as distinctive as the reef face. They may be studded with coral-encrusted boulders, or paved with sand which may itself support a rich blanket of life in the shape of beds of seagrass. Towards their rears the flats may become shallower, and may even have intertidal zones, as a result of dead coral rubble that has been washed up onto the flat.

Coral reef diversity

Like tropical rain forests, coral reefs are extraordinarily diverse, not only in terms of the many different types of coral in every reef, but in the huge range of marine animals that depend on them for shelter and for food. There are over 700 species of reef-building coral in the Indian and Pacific oceans, and in some reefs around the Philippines there may be as many as 400 in a single reef. Diversity tends to fall off away from Southeast Asia, however, and there are 200 off Madagascar and in the Red Sea, around 170 near the eastern side of the Malay Peninsula, 117 off India, 57 in the Arabian Gulf, and only 35 in the Atlantic and Caribbean.

This rich and productive environment provides a wealth of homes and feeding opportunities for numerous creatures. By day, hundreds of different species of fish are active in and around the reef – many of them brilliantly colored to help them find the right mate among the crowd. But the pressure of these voracious fish drives the mobile invertebrates into hiding. Even the corals retract their tentacles. By night, however, the fish disappear, and the corals put out their tentacles to feed on plankton and squids, crabs, mollusks, sea urchins, and many other creatures emerge from their daytime hiding places among the coral.

Coral disturbance

The overall productivity of the reef systems is between 30 and 250 times as great as that of the open ocean. Yet each reef is a distinctly fragile habitat which depends for its high productivity on the recycling of nutrients within the ecosystem as a whole. Its complex food web is very easily disturbed by over-harvesting. Overfishing can lead to dramatic changes in community composition: in some reefs from which grazing fish have been removed, the coral community has become dominated, even threatened, by the subsequent growth of algae.

Oceanographers have estimated that as much as 10 percent of the world's coral reefs have already been degraded beyond recovery, and that a further 30 percent are likely to decline seriously within the next 20 years unless action is taken to protect these sensitive systems.

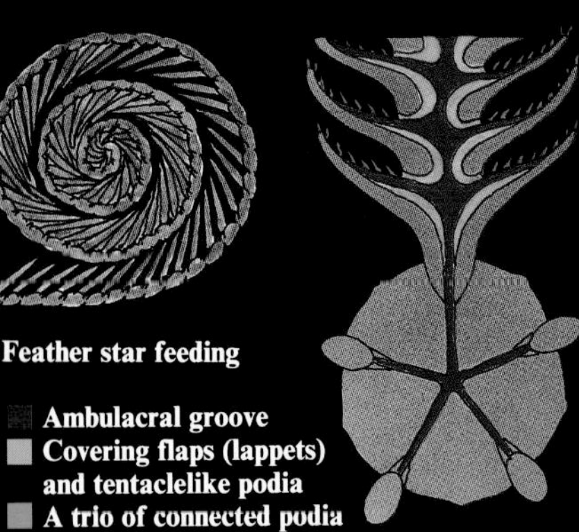

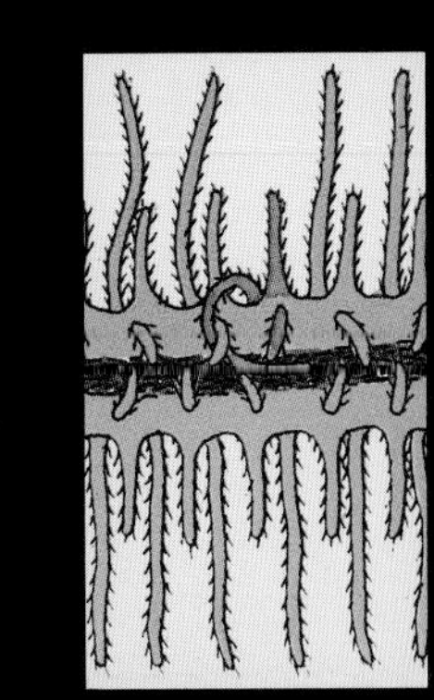

Feather star feeding

- Ambulacral groove
- Covering flaps (lappets) and tentaclelike podia
- A trio of connected podia

FILTER-FEEDING

Filter-feeding is probably the most widespread method of feeding among the sedentary organisms of a reef ecosystem. It requires an abundance of suspended organic particles, phytoplankton, and zooplankton in the ocean waters that pass over the surface of the reef. A typical filter-feeding animal sits attached to the reef, its arms outstretched to form a net that traps the particles and removes them from the passing water current.

Sponges are filter-feeders, as are corals. Surrounding its central mouth, each individual polyp or coral animal has a crown of small tentacles that trap and hold its planktonic food.

The feather star (8E) also has a central mouth, from which branch its arms. Each basic arm may subdivide several times, eventually into small jointed appendages called pinnules. A groove lined with tiny, mobile hairlike projections extends from the mouth right to the end of each pinnule.

This ambulacral groove is protected by lappets, each of which produces a microscopic forest of filtering podia, arranged in groups of three: two long and one short. Special glands on the feather star's tube feet produce the mucus in strands that float close to the surface of the outstretched arms, and that are held in position by the pinnules and podia. Particles in suspension trapped by the mucus net are scraped into the ambulacral groove by the podia, and passed by the ciliary hairs down the arms to the central mouth.

Feather stars are related to the well-known brittle stars, which are much more active animals that crawl over the surface of a reef to pick up small particles of food. The mouth of a feather star faces upward, unlike that of a brittle star, and the feather star mostly remains in one place, though unlike its fixed, deep-water relatives the sea lilies, feather stars can swim if they have to. When food supplies are inadequate or the current becomes too strong, a feather star detaches itself from the reef and swims to a more amenable sight by waving alternate sets of its numerous arms.

PREDATORS OF THE REEF

The largest and perhaps best-known predators of coral reef communities are the sharks which cruise off the reef, feeding on a wide variety of larger marine animals. The ends of the channels through a reef system, through which water from the shoreline can pass out to the open ocean, often serve as assembly points for sharks (18E), which swim up and down the area waiting for fish and other animals to leave the reef as the tide changes.

Tiger sharks even feed on the venomous sea snakes *Laticauda colubrina* (18H), which in turn feeds mainly on predatory eels. The female snakes are much larger than males, and tend to eat the big congrid eels to be found in deeper water off the reef face and in surge channels, whereas the males hunt for moray eels in the reef face, closely inspecting cracks and holes where an eel may be hiding. The snakes' powerful venom is used to subdue the normally aggressive eels, before they are ingested whole.

Unlike many sea snakes which never come ashore, *Laticauda* does come out on to land, and colonies may be found on small reef islands throughout the Indian-West Pacific.

16 | 19 | 20

4 33-65 feet deep

E

F

16 | 17

REEF STRUCTURE

Wave energy is one of the two dominant factors that control the distribution of coral species and the other animals associated with them. (The other factor is light.) The seaward face of a reef environment is exposed to considerable wave energy in the full force of ocean swells. During storms, powerful waves may break off entire coral colonies.

The area that experiences the greatest wave action is the reef crest, the seaward edge of the flattish upper surface of the reef, on which waves may break at low tide, and over which cool ocean water passes at other times. Despite the high level of wave energy, this is the zone of most vigorous coral growth because the cool oceanic water brings plankton to the reef – plankton that feeds the corals and the multitude of filter-feeding animals that live in the reef.

Behind the reef crest, where wave energy is lower, there are generally finely-branched corals such as *Millipora* (7E) and *Stylophora*, often in association with soft corals such as the bowl corals (2C). The upper surface of the reef is also the area where encrusting algae grow most vigorously. Their low-profile growth protects them against disturbance by the waves even as they cement the otherwise loose materials of the reef together.

On the reef face below the reef crest are ranged the *Acropora* (9D), their large, strong branches resisting the force of the breaking waves, and squat forms like the brain corals (10E). Below the zone of high wave energy live more fragile growth forms, such as the sea fans (13E) and the platelike corals (11F). The flattened, upward-facing surfaces of these corals provide a huge habitat in which the zooxanthellae can absorb the sunlight necessary for their photosynthesis. At greater depths neither the reef-building nor the stony corals can survive because there is insufficient light, for the zooxanthellae. Yet even here there are relatives of the reef-building corals that do not possess symbiotic algae, such as black coral (20H).

The reef flat and the reef crest are also the locations of the most vigorous growth by larger macroalgae (seaweeds), which provide food for herbivorous fish and larger sea urchins.

Different species of sea urchin live in different areas of the reef, reflecting the distribution of their algal food and their various capacities to withstand wave action. *Tripneustes gratilla*, for example, occurs in seagrass and algal communities in sheltered lagoons, in which it drapes pieces of weed over itself to hide. The small *Echinometra mathei* grazes on the algae of the reef flat at night and hides in coral during the day. The large slate-pencil urchin (2D) uses its strong spines to wedge itself securely in crevices in the reef face. The spines of *Diadema savignii*, although poisonous, do not protect it against parrotfish (11E) which bite off each spine in turn before flipping the defenseless urchin over to eat it from the underside.

11 | 12 | 13 | 14 | 15

SPONGES

In reef areas where there are high concentrations of suspended sediment, sponges and soft corals may dominate the community. Despite their plantlike appearance, sponges are animals, and feed by drawing water in through inhalant pores and passing it through the body tissues. Particles in the water are trapped inside the sponge, and are absorbed directly into individual cells (much as single-celled protozoans take nourishment).

The variety of body forms in sponges is almost as diverse as that of corals. Upright, branched forms live in areas of low water movement, and massive cone-shaped forms grow in sandy patches of the lagoon and at depth (16H). Instead of the solid skeleton of the corals, however, sponges rely either on small spicules of bone matter embedded in the tissues, or on a matrix of organic fibers around which the soft tissues of the animal are organized. Within the body of the sponge there may be numerous small animals such as tiny crustaceans, brittle stars, and worms, which siphon off food particles from the water drawn into itself by the sponge.

CORAL DYNAMICS

The key to survival

- Symbiosis between coral and photosynthetic algae
- Shallow water that allows photosynthesis to occur
- Consistently high water temperatures
- Well-oxygenated water
- A plentiful supply of microscopic plankton, on which the living coral polyps feed

Forces for change

- Hurricanes and storms
- Outbreaks of predators, such as the crown of thorns starfish, also the attentions of burrowing animals
- Algal overgrowth through a loss of grazers, or through nutrient enrichment
- Coral mining and blast fishing destroying the physical structure
- Sedimentation, smothering and stressing corals
- Coral bleaching as water temperatures rise

Bathed all the time by warm tropical waters, corals are highly sensitive to any change in their stable, comfortable environment, whatever the cause. In recent years, they have been particularly disturbed by human activities, and a considerable percentage of the world's reefs have already been destroyed, or degraded beyond any hope of recovery.

Short-term changes

Although some individual coral colonies can live for decades or even centuries, the ecosystem as a whole is subject to constant change. The population of short-lived species fluctuates rapidly in response to various stimuli. The corals are preyed upon by the crown of thorns starfish *Acanthaster planci*. Coral cover is severely depleted when this starfish undergoes its regular explosive increase in numbers.

Where coral cover is reduced by natural events or human activity, recovery may be slow. If the animals that graze on the symbiotic algae are removed, the algae may choke the coral and the coral may not recover at all. This happened in Jamaica, where overfishing and disease decimated herbivorous fish and sea urchins. After a hurricane destroyed the coral cover in 1980, algae smothered the young corals. Coral cover dropped from 50–70 percent to less than 5 percent.

Corals are also very sensitive to temperature. Yet numerous coral communities are living close to their upper limits of temperature tolerance. Even a slight warming of the sea surface can bleach coral. This happens because the coral extrudes its symbiotic algae and loses its color and in extreme cases dies. There are signs of coral bleaching taking place in the Indian Ocean and the Caribbean.

High levels of suspended sediment in the water can harm corals too. Where dredging or placer mining occurs close to reefs, the coral community may be

Storm succession

Hurricanes dramatically alter the nature of a coral community (above left). Just after the storm (above), the surface of the reef is covered with coral rubble and broken coral colonies. The surface is too unstable for juvenile corals to grow on. Gradually, though, currents sort the fragments and coralline algae grow on their surface cementing them together. As the stability of the surface increases, juvenile corals settle and begin to grow, eventually forming the dense coral gardens characteristic of undisturbed areas. But for a long time, the community is characterized by low-growing, short-lived species (left). The longer-lived species may take decades to reach the same density. Which species re-establishes itself first, though, is often a matter of chance.

smothered or die because of reduced light intensity. An even more disturbing threat to reef survival, however, is the increasing discharge of untreated sewage into inshore waters close to coral reefs. The nutrients in the sewage encourage growth of macroalgae that smother young corals.

Far-reaching effects

Not only the corals, but animals that live on the coral, can be put at risk by changes in the environment. One reason coral reefs are such an incredibly productive biome is because of the efficient way they recycle nutrients. Yet there is a limit to the sustainable harvest from coral reefs, which has been calculated at about 20 tons of carbon per square mile. This is an extremely small figure by commercial agricultural standards; for example, modern rice production techniques can yield over 400 tons. So reef systems are highly susceptible to overfishing. In areas where fish populations are declining, fishermen may resort to destructive fishing techniques to harvest even the smallest fish. Dynamite fishing is a widespread problem from the western Pacific to the east coast of Africa. The shock wave not only kills the fish but shatters the coral skeletons, leaving a desert landscape of broken coral fragments.

Coral houses

In some places, such as the Maldives, coral is the only source of building material. Indeed, the Maldives are entirely made of coral sand on top of an atoll reef. In the past the low population density meant that the harvesting of corals for building materials did not happen on a large enough scale to threaten the survival of the reefs. With the increasing population, however, the rate of harvest has now become unsustainable. There have been attempts to restrict or ban the harvest of coral for building. But in some places, there is no sign of recovery even 50 years after the harvesting of coral has ceased. This is because the flattened surface of the reef is subject to strong current action and the larval corals cannot settle and attach onto the rubble surface.

Reef tourists

A more recent stress to coral reefs is the increasing number of tourists visiting reefs to dive. Divers often alter reefs by collecting shells and breaking off coral, and carelessly cast boat anchors often plow through the fragile coral gardens. Moreover, the construction of tourist resorts not only increases sediment levels around the reef but also ensures that extra sewage will be pumped into the sea. As a result the tourist industry destroys the very resource on which its economic survival depends. The ultimate viability of coral reefs may depend on wise management, and the limiting of human activities that upset the delicate balance of these communities and alter their composition irrecoverably.

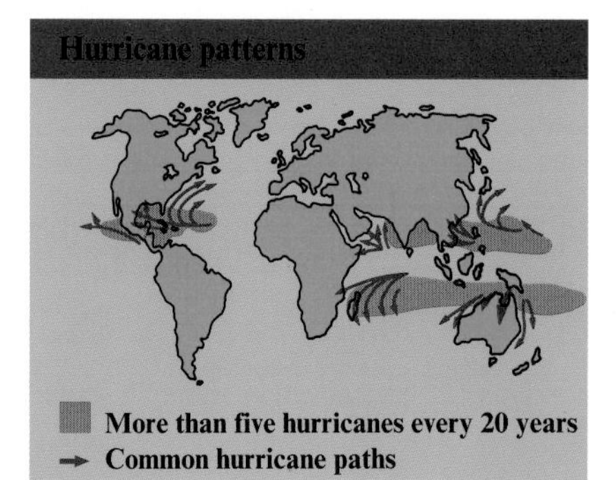

Coral shapes

Corals grow throughout the tropics, but those growing in hurricane-prone areas such as Guam in the Pacific are quite different from those in calmer areas such as the Maldives. Hurricane-prone reefs are generally composed of coral species with a low and compact growth form; finely branched forms are not usually found. In contrast, corals growing on reefs without regular storms tend to display a wide variety of branched and upright species.

Hurricane damage

Known as typhoons in the Pacific, hurricanes wreak havoc whenever they strike delicate coral communities. The ferocious winds that circle the eye of the storm at speeds of up to 200 mph (1) generate mountainous waves and help pile up a huge surge of storm water that locally raises the sea level 18 feet or more. As a hurricane sweeps across a coral island (2), the island is swamped by the storm surge and waves batter the coral communities so violently that many are destroyed. Skeletons of finely branched species growing on the reef are snapped off and platelike corals growing in deeper water on the reef face are ripped clean away. After the storm passes over (3), the waves subside, but the corals are still threatened not only by the deluge of fresh water running into the sea after the torrential rains, but by the sediment and debris washed down with it, all of which slows down the coral communities' recovery.

GLOSSARY

allelopathy The effects that the metabolic products of a plant have on the growth of its neighbors.

altruism Behaviour which aids another organism without benefitting the performing organism.

aquifer A permeable underground rock stratum, capable of holding water.

anoxic Lacking oxygen.

barbel A slender tactile bristle hanging from the lower jaw of a fish, or one of several species of fish possessing such bristles.

benthic Relating to the bottom of a body of water.

bioluminescence The production of light by living organisms.

biomass The dry mass of all living material in an area.

biome A major regional ecological community, usually defined according to the dominant plant types present.

biosphere The layer of the Earth that contains living organisms.

browse The above-ground part of shrubs and trees available for animal consumption.

chamephyte Herbaceous perennials with buds close to the ground.

chasmophyte Plant adapted to life on cliff faces, or in gorges.

climax The stable community at the end of a period of succession, which is capable of maintaining itself so long as the environment remains constant.

commensalism A relationship between two species that benefits one but not the other.

continental drift The shifting of the Earth's crust in response to currents in the molten interior.

convergent evolution The development of similar features in different species living in different areas, but under similar ecological constraints.

coppice A thicket or dense growth of bushes or trees, regularly trimmed back to stumps.

cyanobacteria Photosynthetic "blue-green" micro-organisms.

deciduous Leaves that are shed during a particular season.

decomposition Breakdown of complex organic substances into simpler ones.

delta The flat alluvial area at the mouths of some rivers, where the mainstream splits into several distributaries. A delta may empty into the sea, such as the Mississippi's, or form an inland swamp, such as the Okavango's.

detritivore Organism that feeds on dead organic matter (but not fungi or bacteria, which are usually referred to as decomposers).

dinoflagellate Unicellular alga with a cellulose shell and two whip-like extensions, or flagella, which it uses for locomotion.

dystrophic A brown lake with high humus content, little bottom-dwelling animal life and high oxygen consumption

ecosystem The totality of interactions between a community and its non-living environment.

ecotone The transition zone between two structurally different communities.

endemic Restricted to a given area.

epiphyte Organism that lives on the surface of a plant, but does not draw nourishment from it.

estivation Dormancy in animals during a dry season.

estuary Partly closed bay where the waters of a river meet the sea.

euphotic zone The layer of water from the surface to the maximum depth at which light penetration allows photosynthesis.

eutrophication The nutrient enrichment of a body of water, leading to accelerated plant growth.

evolution The change in characteristics of a population of plants or animals over successive generations.

food chain The flow of energy from one group of feeding organisms to another, beginning with green plants.

food web A number of interconnecting food chains.

geophytes Plants that retreat underground to survive dry periods as bulbs or tubers.

foraminifer An amoeboid protozoan possessing a protective shell.

front The division between two air masses with different origins and different characteristics.

gene A unit of hereditary material that controls some characteristic of an organism.

greenhouse effect The theory that certain gasses, such as carbon dioxide, being released into the Earth's atmosphere are trapping heat that would otherwise be free to radiate out into space, thus leading to global warming.

hemicryptophytes Plants with their growing parts at ground-level.

hibernation Winter dormancy in animals.

host An organism that provides food or some other benefit to another organism of a different species.

hypha A thread or filament found in various branching fungi.

hydrophytes Plants that grow on waterlogged or flooded soil.

Intertropical Convergence Zone A band, roughly over the equator, where north-east and south-east tropical winds converge.

island biogeography The study of community structure and species distribution on islands.

isotherm A line on a map linking points of equal temperature.

lek An area where birds congregate for sexual display and courtship, or the activity of displaying.

littoral The intertidal zone of a sea shore; or the zone relating to the shore of a lake.

marsh Wetland characterized by grassy vegetation such as sedges.

microclimate Climatic conditions on a small scale, usually near or beneath some object, that differ from the general climate of a region.

mycorrhizae Mutually beneficial association of a fungus with the roots of a higher plant.

natural selection Darwin's explanation of the mechanism of evolution, often paraphrased as "survival of the fittest."

neoendemic Recently evolved species, currently restriced to an area but capable of dispersal.

net production The accumulation of biomass in a given period, excluding respiration.

niche The role of an organism in the community.

nutrient Substance required to maintain an organism's normal activity, and any regeneration or growth.

paleoendemic Restricted to an area since prehistory.

oligotrophic A lake with abundant dissolved oxygen but few nutrients.

pelagic The open sea.

perennating organs The growing parts of a plant, especially its buds.

permafrost Permanently frozen soil.

phanerophytes Trees and shrubs.

phenology Study of the seasonal changes in plants and animals.

phytoplankton planktonic plants.

plankton Small, floating or weakly swimming plants and animals, found in salt and fresh water.

radiolarian Protozoan with an often elaborate silica skeleton.

rain shadow Dry area in the lee of a mountain.

resource partitioning Behavioral and feeding patterns that result in the division of the resources of an area between the different species living there, or between males and females of the same species.

ruminant A mammal that chews the cud, or regurgitates its food for further chewing.

sclerophylls Plants with leathery evergreen leaves that combat water loss.

sessile Permanently attached to a substrate, not free to move.

stomata Pores in a leaf through which gases are exchanged.

succession The replacement of one community of plants and animals by another.

symbiosis Two or more species in intimate association.

therophytes Plants that survive unfavorable conditions as seeds.

transpiration The loss of water vapor by land plants.

xerophytes Plants adapted to life in dry or saline environments.

zooplankton Planktonic animals.

zooxanthellae Symbiotic algae.

INDEX

Page numbers in **bold** type refer to major entries; page numbers in ***bold italic*** refer to panoramas; page numbers in *italic* refer to picture captions.

A

abalone 48, ***55***
  Haliotis cracherodii 55
acacias 84, 89, 114
Acacia tortillis 84
actinomycetes 115
albatross 41
alder 18, 26, 58, *79*, 98
algae 41, ***55***, ***56***, 57, 105, 107, 112, 113, ***132***, ***134***
  macroalgae 48
  zooxanthellae ***132***, ***134***
allelopathy ***119***
alligator *59*
alpaca 100, 101
Amazon River 108-11
Anaphalis 97
anchovetta ***44***, 48
Andean mountains ***100-3***
anemone fish ***133***
anemones 14, 75, ***76***
  wood anemone 80
anglerfish ***45***, ***46***
anhinga 59
ant-plant (*Myrmecodia*) ***93***
Antarctica 34, 35, 40, 41
antelopes
  addax 115
  eland 88
  gazelle 14, ***84***, 88
  gerenuk ***84***
  Grants gazelle 88
  impala ***84***
  pronghorn antelope 66
  red lechwe ***63***
  Thompson's gazelle ***87***
  wildebeest ***84***, ***87***, 88
apes 91
Araucaria ***92***
Arctic 34, 40, 41
Arctic poppy ***38***
armadillo ***69***
arowhana ***109***, ***110***
Artemisia 14, *15*
ash 74, ***78***, ***79***, 81
aspen *18*
aster 105
atolls 130
auroch 74, 81
avocet 59

B

baboon *23*
bacteria 14, 15, 19, 43, ***47***, 105
  cyanobacteria *18*
badger *23*, 75, 123
bamboo 98
banksia 122
baobab *83*, ***84***, 88, 89, 114
barbel 106
barnacles 56
  Balanus balanoides 52, ***53***
  Chthamalus stellatus 52
  goose barnacle (*Pollicipes poly merus*) ***55***
bats 22, ***95***
  fruit bat ***93***
bears 75, 81
  brown bear 31, 75
  polar bear 22, 34, ***38***
beaver dams 33
beavers 27, 33, *33*
bedbugs 23
beech 74, ***76***, ***79***, 81
beef cattle 72
bees *22*
beetles ***36***, 91, 106
  ambrosia beetle ***79***
  bark beetle 33
  Bruceid beetles ***84***
  burying beetle ***36***
  deadwood beetles ***78-9***
  dung-beetle ***36***, ***69***, ***124***
  longhorn beetle ***79***
  pinnicate beetle ***117***
  Rhysodes sulcatus 75
  stag beetle ***79***
  Staphylinidae ***78***
begonia ***95***
Bialowieza National Park (Poland) *75*
bilberry 34
bioluminescence ***45***
birch 16, *18*, 26, 27, 34, ***39***, 66, 74, ***76***, 98
birds of paradise ***94***, ***95***
  raggiana ***94***
birthworts ***124***
bison 66, ***70-1***, 72, 74
black poplar 80
black sea dragon ***45***
black-eyed susan ***70***
blackberry 105
blennies 57
  scale-eating blenny (*Aspidontus*) ***133***
bluberry 34, ***37***
bluebell 75, ***76***
bogs 58, *58*, 59, 113
booby ***44***, 48
boreal forest *see* taiga
bower bird 91, 94-5
bream 106
brittlestar ***47***, ***53***, ***55***, ***135***
bromeliads 91, 96
broom ***126***
buddleia *22*
buffalo ***84***, 88
bulrush ***62***
bustard 11, 14
butterflies *22*, ***109***
bird wing butterfly ***95***
  southern festoon ***124***
  speckled wood butterfly ***78***
  swallowtail butterfly ***124***
  white butterfly ***31***
butterfly fish ***133***

C

cacti
  organ pipe cactus ***119***
  prickly pear 11
  saguaro cactus ***119***
caiman 59, 106
Californian coastline ***52-5***
camels 67, ***100***, 115
  bactrian camel ***100***
cameloids ***100-1***
  dromedary ***100***
camouflage ***38***
campions 34, 99, 105
Canadian tundra ***36-9***
capybara 59
cardinal tetra ***110***
carob ***126***
cassowary ***94***, ***95***
cat
  domestic 23
  stray 23
catfish ***60***, ***61***, ***110***
cedar 74
chaffinch ***78***
chamephytes 9
chameleon ***127***
chamise 122
chamois 99
chaparral 122, 123
chasmophytes 128
cheetah 82, ***84-5***, 88, 89
Chernobyl 40
chestnut 98
Chiasmodon niger ***46***
chinchilla ***101***
chipmunks ***30***, 75
chiton (*Katharina tunica*) ***55***
chough 99
cicada 123
cinquefoils 99
Cirolana ***54***
cistus (rock rose) ***126***
clams 57, ***132***
cleaner wrasse (*Labroides dimidiatus*) ***133***
coastlines **50-7**
cockroach 23
coconut 20
collared peccary 123
competition **16-17**
condor 99, 103
coniferous forest *18*, 19, 26, ***30***
continental drift 11
coot 107
coppices 81
Coprosma 97
coral reefs **130-7**
corals ***54***, ***135***
  Acropora ***134***
  Astrangia lajollaensis ***54***
  Balanophyllia elegans ***54***
  black coral ***134***
  bowl corals ***134***
  brain corals ***134***
  Millipora ***134***
  Stylophora 131, ***134***
cormorant ***44***, 48
cotoneaster (*Cotoneaster microphyllus*) *22*
cottonwood 67
cowbird 73
coyote 22, ***68***, ***69***
coypu 59, 64
crabs 51, ***56***, 96
  decorator crab (*Loxorhynchus crispatus*) ***55***
  European shore crab 51
  fiddler crab 51
  hermit crab (*Pagurus samuelis*) ***53***
  horseshoe crab 57
  Petrolisthes ***54***
  rock crab (*Pachygrapsus crassipes*) ***53***
cranberry ***37***
crane 67
creosote bush 115
cricket 22, 23
crocodiles 59, ***62***, 106
crossbills ***31***
  common crossbill ***31***
  parrot crossbill ***31***
  two-barred crossbill ***31***
crow 22
crowberry 34
cuscus ***93***
cyclamen 122
cypresses 123
  Lawson cypress 91
  swamp cypress 58

D

daisy (*Asteraceae*) 21
Darwin, Charles 12, 13, ***30***
deadwood 75, ***78***, ***79***, 81
Death Valley ***119***
deciduous forest 74
deer 27, ***79***, 82, 91, 105
  caribou ***36***, ***37***, 40
  elk 27
  fallow deer ***78***
  pampas deer *83*
  red deer 75, ***78***
  roe deer ***28***, ***78***, 123
  whitetail deer ***71***
defoliating insects 33
desertification 120-1
deserts **114-21**
discus fish ***108***
dodo 21
dolphins ***44***, ***109***, ***110***
  Amazon River dolphin ***109***
  estuarine dolphin ***109***
Draba oligosperma ***103***
dragonfly 58, ***61***
ducks 106
  shelduck 13
  torrent duck ***102***
duckweed 107
duiker 123
dunlin 13
Dutchman's pipe (*Aristolochia*) ***95***

E

eagles ***39***, ***71***, 99
  African fish eagle ***62***
  golden eagle 14, ***116***
  harpy eagle 91
  imperial eagle ***126***
  monkey-eating eagle 91
East African savanna ***84-7***, 89
eastern phoebe 10-11
echidna ***94***
ecosystems
  biomass 14, 19
  biomes 8, 9
  energy-gathering capacities 24
  exploitation of 15
  food webs 14
  mature ecosystems 19
  recycling of resources **14-15**
  succession **18-19**
  types of **8-9**
Edelweiss 99
eels 106, 135
  gulper eel ***46***
  moray eel ***133***, ***135***
El Niño 48, *48*, 96, 97, 121
elephant 21, ***61***, ***84***, 89, *89*, 91, 106
elm ***79***, 80
emu 83, *83*
epiphytes 90, 91, ***92***, ***93***, ***94***, 96
erosion 56
eucalypt *83*
Eucalyptus deglupta 96
eutrophication 107
Everglades, The 59, *59*
everlasting flower 105
evolutionary theory 12

F

fecal rain ***47***
falcons 39, 99
  Eleonora's falcon ***127***
  hobby *77*
  kestrel 22
  merlin ***39***
fan-palm ***126***
feather-star ***132***, ***135***
ferns 74, 75, 91, ***92***, 97, 98, 106
  bracken fern 105
  Cyathea 97
  parsley fern 105
  Pyrrosia 92
  water-fern (*Salvinia*) 11
filter feeding ***47***, 135
finches *12*, 13, 75
  ground finch (*Geospiza fortis*) *12*, 13
  snow finch 99
  vampire finch (*Geospiza difficilis*) *12*, *13*
  woodpecker finch (*Cactospiza pallida*) *12*, 13
fireweed (*Epilobium angustifolium*) 105
firs 26
fishing industry 40, 41, 49, 137
flamingoes 13, ***100***, ***101-2***
  Andean flamingo ***101***, ***102***
  Chilean flamingo ***101***, ***102***
  James's flamingo ***101***, ***102***
flashlight fish 45
fleas 23

flies
- alderfly 106
- blackfly 37, 106, 112
- blowfly ***36***
- caddis fly 106
- mayfly 106
- pine sawfly 33
- stonefly 106

flycatchers ***71***
- collared flycatcher 12
- pied flycatcher 12

foxes 14, ***36***, 75
- Arctic fox ***38***
- fennec fox 115, ***117***
- kit fox ***117***
- mountain fox ***102***
- red fox 22-3

French lavender ***126***, 128, 129
fritillaries 14, 122
frogs
- marsupial frog 96, *96*
- reed frog ***61***
- tree frog 96

fungi ***78-9***, 81, 115
fur trade 32
fynbos 122, *123*, 129

G

Galapagos Islands 13, 49
garbage feeders 23
garigue 123
geese 67
- Canada goose ***71***

gentian 99
geophytes 9
gharial *59*
giraffe ***84***, 88, 89
glaciers *18*, 27
global warming 104
gnats ***31***
goats 104, 123, ***124***, ***125***, 128
goat's rue *22*
Gobi desert 114
godwit 13
goldenrod 22
gopher 105
gorilla *10*, 91
grape hyacinth 122
grasses
- blue grama grass ***70***
- bluestem grass ***70***
- buffalo-grass ***70***
- couch grass (*Elytrigia juncea*) 18, 19
- feather grass 66
- fescue 66, ***70***
- hippo grass ***62***
- Indian grass ***70***
- marram grass (*Ammophila arenaria*) 18, 19
- needle-grass ***70***
- *Pennisetum* 87
- prairie grasses ***70***
- red oat grass 87
- savanna grasses 87
- saw grass 58
- switch-grass ***70***
- wheat-grass ***70***

grayling 106
Great American Desert ***116-19***
Great Basin of Nevada ***116***, ***117***, ***118***
Great Kavir desert 14, 15
grebe 106
greenfinch 22
greenhouse effect 41
groundsel 98
grouse 27
- capercaillie 27
- sandgrouse 14
- willow grouse ***38***

grunion 51
guanaco ***100***, ***101***
guillemot 34
Gulf Stream 43
gulls 22, 44, 57
- black-headed gull 23
- herring-gull 22, 23

H

hackberry 67
hakea 122
hares
- Arctic hare ***38***
- brown hare ***124***, ***125***
- snowshoe hare 39

harvester ant ***116***
hatchetfish ***46***
Hawaiian islands 21
hawks ***68***, ***93***
- goshawk ***31***
- sparrowhawk *77*

hazel 74, ***78***, ***79***, 81
heathers 91, ***126***
hedgehog 75, *77*
hemicryptophytes 9
hemlock 74
herb paris 80
herbs ***76***, ***77***, 98, 123, ***125***, 128
heron 59, 106
hibernation ***30-1***
Himalayas 98, *99*
hippopotamus 59, ***62***, 106
holly 75, ***78***
hornbill 91
horses 12
- wild horse ***117***, ***118***

hurricanes 137
hydrophytes 58
hyena 22, ***84***, ***86***
hyrax 123

I

ibex 99
ibis 59
ice sheets 34
iguana 49
Indian blanket ***70***
Indo-Pacific reefs ***132-5***
Intertropical Convergence Zone (ITCZ) 121
ironwood 114
islands **20-1**
ivy 75, 80

J

jack rabbit ***119***, 123
jackal 22, 67
jaguar 91, ***102***
Japanese knotweed *22*
jay 22, ***30***, ***76***
jerboa 115
Jerusalem sage ***126***
juniper tree 98, 128

K

kakapo *21*
kangaroos 82, 123
- tree kangaroo ***93***

kapok tree ***111***
kelp 57, *57*
killifish ***109***
kingfisher ***63***
kite 59
kiwi *21*
klinki tree 91
knife fish ***110***
Krakatau 21
krill (*Euphasia superba*) 41

L

laburnum *22*
Lake Baikal 107
lakes *see* rivers and lakes
lammergaier 99
lanternfish ***46***
larch 26, 27, ***28***
lark 14
Larrea 114
Lauraceae ***95***
laurel 98
leafhopper ***86***
leech 106
lemming ***29***, ***36***, ***38***
lemur 91
lentisk 128
leopards 82, ***84***
- clouded leopard 91
- snow leopard 99

lice 23
lichens *18*, 26, ***37***, 40, 81, 91, ***92***, 98
- *Cladonia* ***37***

lime 74, 80, 81
limpets 51, ***53***, 56, 57
- *Acmaea digitalis* ***53***
- *Acmaea persona* ***53***
- keyhole limpet (*Diadora aspera*) ***55***

linnet *22*
lion ***61***, 82, ***84***, ***86***, 123
liverwort ***92***, 106
lizards 96, ***100***, ***102***, 115, ***117***, 123, ***124***
- chuckwalla ***117***
- fence lizard ***119***
- gila monster ***117***
- green lizard ***127***
- Komodo dragon 21
- *Liolaemus multiformis* ***102***
- Salvadori's monitor ***94***
- slow-worm 22
- thorny devil 115

llama ***100***, ***101***
lobelia 98
locust ***86***
logging 32, 81
loris 91
lungfish 58
lupin 105
lynx 27, ***28***, ***39***, ***71***, 123

M

magnolia 10, 11, 98
magpie 22
mallee 122
manatee ***111***
mangrove swamps 58, 59, 64, 65
mangroves 58, 64-5, 96
- black mangrove (*Avicennia germinans*) 65
- button mangrove (*Conocarpus erecta*) 65
- red mangrove (*Rhizophora mangle*) 65

maple 74, 98
maquis ***124-7***, 128, 129
marmot ***30***, 99, 104, 105
marsh harrier 59
matorral 122
mesquite (*Prosopis juliflora*) 88
mice 22, ***118***
- brush mouse 10

Michaelmas daisy 22
midges ***31***, 35
migration ***60-1***
mining 32
mist forests 91
mites *77*
Mkomazi Game Reserve 89
Mojave desert *17*
mole 75
mole rat 115
mollusks ***47***, 51, 107, ***132***
- *see also* individual species

monkeys 91
- spider monkey 91, ***111***

moorhen 107
mosquitoes 31, 37
- *Aedes aegypti* 23

mosses *18*, 19, 26, 37, 74, 75, 81, 91, 92, 98, 106
- sphagnum moss 58, 64

moths
- elephant hawkmoth 22
- yucca moth 17

mountain sorrel *18*
mountains **98-105**
mudskipper 64
mushrooms 78
musk ox 36
mussels 55, 56, 106
Mytilus californianus 54
mustang 118
mycorrhizae 77-8, 91
mynah 23
Myristica 95

N

natural selection **12-13**
New Forest ***76-9***
New Guinea rain forest ***92-5***
nightjar 91
North American dustbowl 73
North American prairie ***68-71***
Nothofagus 91, ***93***
nutcracker ***30***, ***31***
nutmeg ***92***

O

oaks 12, 66, 74, ***76***, ***78***, 81, 91, 98
- holm oak 123, ***126***
- kermes oak 126, 128
- white oak 123

oceans **42-9**
ocelot 91
Ogallala aquifer ***68-9***, 73
oil exploration and extraction 40
Okavango Delta ***60-3***
olive 123, 128
opossum *23*, 91
orchids 91, ***92***
- bee orchid ***126***
- fly orchid ***126***
- spider orchid ***126***
- wasp orchid ***126***

oriole 73
ostrich 83, *83*
otters 106
- giant otter ***111***
- sea otter 57, *57*

overgrazing 88-9, 104-5, 120-1
owls
- burrowing owl ***68***
- great grey owl ***29***
- Pel's fishing owl ***62***

oxbow lakes 113, *113*
oxpecker ***63***
ozone layer 41, 104

P

palm tree 10
palo verde 114
paperbark tree 83
Papuacedrus 91
papyrus 58, ***62***
parasites 16, ***94***
Parkia ***93***
parrotfish ***133***, ***134***
peat bogs 58, 64
pelicans 71
- brown pelican ***44***, 48

pink-backed pelican ***63***
penguins 34, 41
- emperor penguin 35
- nesting penguins *17*

perch 106, 107
periwinkles 57
- *Littorina planaxis* ***53***
- *Littorina scutulata* ***53***

permafrost ***28***, 40
photosynthesis 14, ***44***, ***95***
pigeon 22, ***94***, ***95***
pike 106, 107, 113
pill bug ***52***
pines *18*, 26, 66, 74, 123, 128, 129
- Aleppo pine 128
- Calabrian pine 128
- hoop pine 91
- Jack pine 27

kauri pine 91
Scots pine *18*, 27
Siberian stone pine 27
pingoes ***39***
piranha 106, ***110***
plankton 48, 49, 112-13, ***134***, ***135***
Gonyaulax ***54***
phytoplankton 41, *42*, 42-3, ***44***, 48, 49, 107, 112, 113, ***135***
zooplankton ***44***, 107, 112, 113, ***132***
plover 51, 57
polar regions **34-41**
pollution
acid rain 32-3, 64
coastlines 57
mountains 105
oceans 49
rivers and lakes 113
taiga 32-3
tundra 40
wetlands 64
polygons ***39***
pondweed 107, 112
red pondweed (*Potamogeton alpi nus*) 107
prairie **66-73**
prairie chicken ***69***
prairie dog ***68***
prairie iron weed ***70***
predation **16-17**
protea 122, 123
ptarmigan ***39***, 99
puma ***102***, 123

Q

quail ***116***
quelea ***86***
quoll ***94***

R

rabbits 11, 82, ***124***, ***125***
cottontail ***71***
raccoon 23
Rafflesia ***94***
rain forest **90-7**
Ranunculus 97
rats
cotton rat ***71***
kangaroo rat 115, ***117***, ***118***, 123
muskrat 67
Norway (Brown) rat 22
Raunkiaer, Christen 8, 9
redshank 13
redstart 22
redwood 74
reed warbler 107
reeds 62, 107, 113
resource partitioning 17, ***118***
rhea 82, 83, *83*
rhinoceros 89
rhododendron 23, *97*, 98
rivers and lakes **106-13**
roach 107
rose root 105
rosebay willowherb 22
rosewood ***92***, ***95***
rotifers 112
rubythroat ***31***
rudd 107
ruminants ***124***
rushes 107, 112
Russian taiga ***28-31***

S

sagebrush 66, ***119***, 122-3
Sahara 114, 120, 121
Sahel 120-1
salamander 81
salinity 43
saltbrush (*Atriplex confertifolia*) ***118***
sand plum tree 67
sardine ***44***, 48
Sargasso Sea 43, *43*
savanna **82-9**
saxifrages 34
scallop 48
Sciaena deliciosa 48
scilla 75
sclerophylls 9, 122, ***125***
scorpion 115, ***116***, ***119***
scrublands **122-9**
sea anemones ***54***, 56, ***133***
Anthopleura elegantissima ***54***
Anthopleura xanthogrammica ***55***
Epiactis polifera ***54***
sea cucumber ***47***
sea fan 132, 134
sea lily ***47***
sea pen ***47***
sea slug ***55***
sea spider (*Pygnogonum stearnsi*) ***55***
sea stars *see* starfish
sea turtle 57
sea urchins ***53***, *56*, 57, ***134***
Diadema savignii ***134***
Echinometra mathei ***134***
slate-pencil urchin ***134***
Tripneustes gratilla ***134***
sea whip ***132***
seals 34, 35, 56
fur seal 41
seamounts 43
seaweeds 55
bladder wrack 50
Cladophora ***55***
Iridaea ***55***
kelps (*Laminaria*) ***52***, ***55***, 56, 57
knotted wrack 50
Pelvetia 53, ***55***
secretary bird 83
sedges ***37***, 58
Serengeti National Park ***87***, 88
sewage discharge 49
sharks ***44***, ***135***
cookie-cutter shark ***46***
tiger shark ***135***
sheep ***124***, 128
domestic 104-5
mountain ***101***, 104
pronghorn sheep ***71***
shrimp 41, 106
Silene acaulis ***103***
snails ***61***
sea-snail (*Opalia crenimarginata*) ***55***
water snail 112
snakes 91, 96, ***102***, 115, 123
anaconda ***111***
cat-eyed snake ***61***
Montpellier snake ***127***
python ***93***
rattlesnake ***68***, ***117***
sea-snake (*Laticauda colubrina*) ***135***
smooth snake ***127***
vipers ***117***
water-snake 67
Southeast Pacific ***44-7***
sparrow 22, 73
species
competition **16-17**
distribution patterns **10-11**
natural selection **12-13**
neoendemic 10
paleoendemic 10
predation **16-17**
symbiotic relationships 17
spiders ***36***, 115, ***116***
black widow spider ***116***
tarantula ***116***
trap-door spider ***116***
sponges ***47***, ***132***, ***134***, ***135***
spoonbill ***63***
springtail ***36***
spruce *18*, 26
black spruce 27
Norway spruce 18
white spruce 27
squid ***46***
squirrels ***30***, 75, ***79***, 105, 115, 123
flying squirrel 81
grey squirrel 11
hibernation ***30-1***
red squirrel ***30***
starfish ***53***, ***55***
Acanthaster planci 136
Pisaster ochraceus 54
starling 22, 73
steppe 11, 66, *67*
stick insect ***95***
stilt 13
stingray ***108***, ***110***
stone loach 106
stonecrop 99
storks
Abdim's stork ***61***
African stork 83
Maribou stork *23*
strawberry tree ***126***
succession **18-9**
sugar glider ***93***
sunflower 66
swamps 58, 59, ***60-3***
swift 22
swordfish ***46***
sycamore *22*, 23
symbiotic relationships 17, ***132***, ***133***

T

taiga **26-33**
takahe *21*
tamarisk 114
tapir 91, 106
tarsier 91
taun (*Pometia pinnata*) ***92***, ***95***
temperate forests **74-81**
temperate grasslands **66-73**
tench 107
Terminalia ***95***, 96
termites ***63***, ***78***, ***86***
terns
Arctic tern ***38-9***
least tern 57
Tetramales 95
Thais 52, 54
therophytes 9
thorn tree 83
tick 16
tickweed *17*
tides 50, 51
tiger fish ***60***
tinamou ***103***
tits ***79***
toads 100
horned toad ***69***
Puna toad ***102***
spadefoot toad ***119***
tortoises ***124***
Hermann's tortoise ***127***
toucan 91
tree spurge ***125***
Trematomus 35
tripod fish (*Bathyterus*) 47
trout 106
truffles 78
Tsavo National Park 89
tundra **34-41**

U

urban ecosystems **22-3**

V

vicuna 100-1
vines 123
viper fish 46
vireos
black-capped vireo 73
red-eyed vireo *17*
viscacha 99, 101
volcanic eruptions *21*, *104*, 105, *105*
vole 27, ***29***, 104, 106, 107
vultures ***84***, ***85***, 88
black vulture *23*
griffon vulture 14, ***126***

W

wallaby 82
wallows 71
walrus 38
water lilies 58, 62, 107, 112, 113
yellow water lily (*Nuphar lutea*) 107
water skater (*Asellus*) 112
water-fleas 112, 113
weevil ***36***
wetlands **58-65**
whales 34
blue whale 41
killer whale ***44***
right whale 40
sei whale ***44***
sperm whale ***46***
whelk 56, 57
whip scorpion ***116***
whortleberry *18*
wild ass 14
wild cat 75, 123
wild dog 67, ***84***, ***86***
wild olive tree 126
wild pig 75
wild tulip 14
wild turkey ***71***
willow 18, 26, 34, ***37-8***, ***39***, 58, 98, 113
winkle 56
wolf 27, ***29***, ***36***, 40, ***71***, 75, 81, 123
wolverine 27, ***28-9***
wood wasp 33
woodchat shrike ***127***
woodcock ***79***
woodlouse ***77***
woodpeckers ***79***, 81
Arizona woodpecker 17
black woodpecker ***28-30***
gila woodpecker ***116***
ivory-billed woodpecker 75
red-cockaded woodpecker 75
white-backed woodpecker 75
worms 43, ***47***, 107, ***132***
Arctonoe vittata ***55***
bloodworm 106
earthworm ***77***
nereid worm ***54***
scale worms ***55***
swamp worm (*Alma emini*) ***61***
wormwood *22*

X

xerophytes ***119***

Y

yak ***100***
yew 75

Z

zebra ***87***
zoogeographic regions *11*
Zygophyllum 14, *15*

PICTURE CREDITS

Panoramas

Taiga
SURVIVAL ANGLIA Wothe 1/2A-D, Bailey 17-20A-D, Wothe 6D, Wothe 9-11A-C
NHPA Danegger 16-18A-C, Blossom 12/13C/D, Wendler 1/2C/D, Danegger 14/15C, Zepf 20C/D, Pott 5C, Mutomaaki gutter/D, Tidman 14C
FLPA Hosking 20A, Hosking 11 A/B, Hosking 4D, Whittaker 5 C/D, Wilmshurst 6/7 B/C, Hosking 3D, Newman 10C, Hosking 9A, Willmshurst 99D, Hautala 13 A-D, Newman 1-3A, Wisniewski 4-6AB

Tundra
SURVIVAL ANGLIA Bennett 3-5B, Foott 11D, Steenman 2B, Wothe 11B, Foott 18C, Bennett 12-15C, Steenman 9C, Kemp gutter/D, Bennett 12/13A, Kemp 11D, Kemp 1-3B, Bennett 3-5B, Bennett 17-20A/B
NHPA Alexander 17D, Shaw 8C/D, Danegger 5D, Gainsburgh 6/7D, Krasemann 4D, Krasemann 10C, Hawkes 7/8B, Shaw 5D
FLPA Hosking 14B, McCutcheon 12A, Hosking 17B, Hamblin 16A, McCutcheon 5/6C, Newman 7-10D, Hosking 1-3C/D

Oceans
NHPA Wu 6/7D/E, Agence Natur 19/20F/G, Agence Natur 6-8C/D, Aitken 4/5B/C, Wu 1/2D, Wu 17/18E/F, Wu 11/12D, Wu 12/13E/F, Wu 10E, Agenc Natur 9/10C, Wu 9/10B/C, Wu 8/9D/E, Wu 8/9A/B
FLPA Hamblin 1/3A, Earthviews 5/6 B/C, Hosking 4/5A, Wilson 1B, Wilson 16/17G, Wilson 17/18F/G

Coastlines
SURVIVAL ANGLIA Foott 19/20G, Foott 14/15 D, Foott 3B, Foott 3/4C, Palmer 11/12E/F, Foott 8/9D, Foott 12E, Foott 19G, Foott 17G, Foott 20F/G, Foott 7E, Foott 9/10D/E, Foott 6D, Foott 8E, Foott 6E, Foott 3/4D, Foott 13/14E, Foott 13F
NHPA Callow 1D, Krasemann 12F, Erwin 11F, Wu 11F, Wu 18F, Wu 16/17F, Wu 8E, Wu 7-10B, Papazian 13-15D, Wu 16G
FLPA Wilson 18/19G, Dembinsky Jones 9E, Wilson 20G, McCutcheon D5, Wilson 9D, Wilson 3C/D, Wilson 3D, Wilson 10E, Dembinsky 1-5 A/B, Dembinsky 6C

Swamps, Bogs and Mangroves
SURVIVAL ANGLIA Heald 9/10B, Plage 12/13A, Deeble/Stone 3/4D
NHPA Bannister 5/5D, Bannister 6/7D, Bannister 7/8A/B, Bannister 10/gutter, Johnson 4/5A/B, Johnson B6, Johnson gutter/12B, Bannister 1/2C/D, Dennis 1/2A, Bannister 4-6D, Johnson 3B
FLPA Perry 2C, Haynes 11/12 C/D, Gore 12B, Hosking 10A, Tidman 15/16A/B, Newman 14C, Hosking 8/9C/D

Temperate Grasslands
SURVIVAL ANGLIA Plage gutter/11, Foott 2/3C/D, Foott 1-6A/B
NHPA Tidman 10/11A-D, Erwin 8/9D, Shaw 12/13A/B, Krasemann 18-20B-D, Krasemann 5D, Papazian 9/10D, Krasemann 17/18B/C, Shaw 16-20A/B, Shaw 12/13C/D, Kirchner 6/7D
FLPA Brandl 5D, Lee Rue 9B/C, Brandl 7/8A, Dembinsky 7C, Lee Rue 14/15B/C, Lee Rue 10C, Dembinsky gutter/11D, Lee Rue 8-11A/B, West 16C, Lee Rue 14/15D, Dembinsky 1D

Temperate Forests
NHPA Campbell 12D, Campbell 8D, Tidman 18D, Bainn 13-15D, Callow 8/9D, Paton 6D, Gatwood 8C, Dalton 10D, Shaw 7D, Danegger 11-12C, Campbell 17D, Zepf 10A, Campbell 17-20A/C
FLPA Thomas 1/2D, Withers 5A-D, Hosking 11/12D, Wilmshurst 1-3A-D, Wilmshurst 4A-D, Austing 14A, Hosking 20D, Hosking 3/4D, Hamblin14/15C/D, Batten 7C, Wilmshurst 11D, Lawrence 4/5D/B, Bird 16A-D, Hosking 10C, Hosking 10B, Wilmshurst 7/8A/B, Wilmshurst gutter 12A/B, Walker 12-14A/B.

Savanna
NHPA Johnson 13/14D, Dennis 10/11A, Krasemann 16-20A-D, Shaw 14 B/C, Robinson 4/5C/D. Robinson 6/7C/D, Krasemann 9/10B, Dalton 15/16B-D, Bannister 14B/C
SURVIVAL Deeble/Stone 11B/C, Matthews/Purdy 7B/C, Tibbles 10B/C, Foott 12C, Matthews/Purdy 10C, Bartlett gutter/11 B/C, Root 1/2 A-D, Root 1-20/background.
FLPA Silvestris 8/9D/E, Wisniewki 11B

Rainforest
SURVIVAL ANGLIA Root 1/2B-D, Bray 1-3A-C, Brett 16/17E
NHPA Keller 18 F/G, Newman 2/3 A/B, Keller 3A, Mackay 16/17G, Paton 6D, Watts 5C, German 5/6B, ANT 18/19E/F, Pollitt 3B/C, Mackay 16F/G, German gutter-11B, Keller 20E, Polunin 16E, Surtak 13E, Dickins 16C, ANT 9B-E, Bernard 14 B-D, Krumins 8D/E, Keller 7B-D, Borman 3/4A-D
FLPA Preston-Mafham 18G, Holmes 19/20G, Newman 12/13B/C, Preston-Mafham gutter/C/D, Preston-Mafham 15E, Hosking 11B.

Mountains
SURVIVAL ANGLIA Koster 7/8F, Foott 16C, Huggins 18/19C/D, Huggins 11B
NHPA Gainsburgh 16/17DE, Dixon 18/19A/B, Scott 1/22E-G, Dalton9/10E, Krasemann 11-13B, Bernard 8-10D, Scott 1G, Shaw 2F/G, Silvestris 8C, Krasemann 11-13C/D, Woodfall 3-5D, Bernard 17C, Gainsburgh 15E, ANT 16/17B/C

Rivers
NHPA Palo 16/17A, Wendler 2D, Wendler 16/17C/D, Sauvenet 19B-D, Wendleer 17/18D, Wendler 3-5A-D, Tweedie 1D, Bernard 1-4 A/D, Schafer 3-4A, Wendler 3D, Dalton gutter/11D/E, Wendler 7/8D/E, Wu 6-11B/E, Lacz 15C, Wu 9/10D, Wendler 8E, Sauvenet 6E, Olsen 11/12E, Bannister 12D/E
FLPA Hosking 16-20C/D, Austing 12/13C, Lane 10E, Lane 8D, Lane 12D, Lane 15E, Lane gutter E

Deserts
SURVIVAL ANGLIA Price 17A, Foott 6-8C/D, Linely 16D, Foott 18D, Foott 4D/E, Steelman 1-3A, Price 12D, Foott 14D, Scholfield 6D
NHPA Kraseman 1C, Erwin 20A, Switak 2A/B, Switak 3/4A-C Flanck 17/18B, Switak 17B, Switak 16B, Schafer 15C, Shaw 18-20B/C, Heuclin 19/20C/D, Dennis 1/2C/D, Currey 12-16A, Switak 9/10C/D, Currey 19A, Switak 11D, Shaw 6C
FLPA Newman 17/18D/E, Mc Cutcheon 18A, Lee Rue 6C, Kinzler 7/8A/B, Newman 15/16B/C
ZEFA Bauer 5D

Scrublands
NHPA Heuclin 15/16D, Conseco 11D, Heuclin 1/2B/C, Tidman 7/8A/B, Conseco 13D, Sodeer 6/7C, Meech 16/17B, Van Ingeen 12C
FLPA Borrell 1D, Hawkins 10A, Borrell 19/20D, Wilmshurst 12/13A-D, Wharton 16D, Reynolds 14C, Hosking 18/19A, Tidman 1-5A/B, Hosking 10-13A-C, Tidman 9/10A-C

Corals
SURVIVAL Wills 1/2D, Wills 20F/G, Abbott 1A/B, Plage 3/4C/D, Wills 2C, Pitts 6/7BC, Foott 6/7E, Pitts 13/14F/G, Abbott 2B, Wills 1/2B/C, Abbott 8/9B/C, Abbott 9D, Flood 10E, Wills 8/9B, Wills 13E, Wills 11/12E
NHPA ANT 16G, Wu 18/19G, Wood 8/9E
FLPA Silvestris 16-17F, Carvalho 8/9E, Wilson 14E, Silvestris 11F Silvestris 17-20C/D

Introductions and Dynamics pages

1 FLPA/Earthviews
2 ZEFA
3 ZEFA
4 Hutchinson/Regent
7 FLPA/Newman
9 SPL/NASA
10 FLPA/Ward
13L SURVIVAL/Koster
13R SURVIVAL
15T FLPA/Tidman
15C FLPA/Hosking
15R FLPA/Hosking
16 NHPA/Bannister
16/17 SURVIVAL/A Price
17T NHPA/Shaw
17B NHPA/Switak
18L FLPA/Polking
18R NHPA/Hawks
19 FLPA/McCutcheon
20 ZEFA/Takano
21L NHPA/ANT
21R NHPA/ANT
21BC SURVIVAL/Plage
23L FLPA/Perry
23TR B Coleman/Price
23BR B Coleman/Lockwood
27T SURVIVAL/Rick Price
27BL NHPA/Lacz
27BR SURVIVAL/Tracey
32R FLPA/Earthviews
35L SURVIVAL/Catton
35R NHPA/Krasemann
40T FLPA/Newman
40/41 SURVIVAL/Anderson
41T SURVIVAL/Foott
42 SPL/Feldman, NASA
42ins SPL/Feldman, NASA
43L NHPA/N Wu
43TR NHPA/Carmichael
43BR SEAPHOT/Hessler
48T SPL/Legeckis
48B SPL/Legeckis
49 FLPA/Wisniewski
51 SURVIVAL/Valla
56 FLPA/Wisniewski
57TL SURVIVAL/Foott
57TR FLPA/Lee Rue
57B NHPA/N Wu
59T ZEFA
59C PLANET EARTH/Clay
59B FLPA/Clark
64 SURVIVAL/Plage
65L FLPA/Hosking
65R FLPA/Wilson
67T NHPA/Shaw
67B SURVIVAL/Bennett
72 ZEFA/Photri
73T ZEFA
73B ZEFA/Maroon
74TL ZEFA/Thonig
74TR NHPA/Woodfall
74B SURVIVAL/Bomford
81TL FLPA/Wilmhurst
81TR FLPA/Dembinsky
81 insect L FLPA/Dembinsky
81 insect C FLPA/Dembinsky
81 insect R FLPA/Dembinsky
83TR SURVIVAL/Plage
83CR NHPA/Bachman
83BR SURVIVAL/Price
83BL2 NHPA/Fagot
83BL3 NHPA/Sauvenet
88 NHPA/Shaw
89L FLPA/Hosking
89R NHPA/Robinson
90 FLPA/Hosking
91L NHPA/ANT/Parker
91TR NHPA/Palo
91TC FLPA/Hosking
91BR SURVIVAL/Foott
96 OSF/Fogden
97T FLPA/Premaphotos
97C NHPA/Callow
97R NHPA/Beekles
99 FLPA/Muller
104 ZEFA/Lewis
105TL FLPA/USDA
105TR B Coleman/Gunnar
105B SURVIVAL/Foott
107 NHPA/ANT/Mead
112 NHPA/Wendler
113 NHPA/Wendler
115 NHPA/Dennis
120 SURVIVAL/Davies
121T NHPA/Dennis
121C NHPA/Blossom
121B SURVIVAL/Bartlett
123 SURVIVAL/Park
128 SURVIVAL
129T FLPA/Wisniewski
129B OSF/Fogden
131 ZEFA/Weck
136TL SURVIVAL/Willis
136TR NHPA/Walsh
136B SURVIVAL/Abbott

Abbreviations

T Top
B Bottom
L Left
C Center
R Right
ins inset
NHPA Natural History Photographic Agency
FLPA Frank Lane Picture Agency
OSF Oxford Scientific Films
SURVIVAL Survival Anglia
SPL Science Photo Library

CONTRIBUTORS

Burkhard Bilger is Senior editor of *Earthwatch*, a bimonthly magazine of field science and the environment, and an award winning author

Craig Downer is conducting research at the Centre of Arid Zone Studies, University of Wales, Bangor

Jonathan Elphick is a natural history author, editor and consultant with a special interest in ornithology. He is currently working on a major new multi-media project for the BBC

Dr Mark Everard is a writer and naturalist, working for the National Rivers Authority of Great Britain

Dr Oliver Gilbert is a lecturer in the Department of Landscape, University of Sheffield

Professor Bob Johns is a botanist, attached to the Royal Botanic Gardens, Kew

Professor Cedric Milner is former head of Bangor Ecological Research Station. He has traveled widely in Africa and North America and has paricular interests in conservation and wildlife management

Dr Peter D. Moore is a lecturer in the School of life, Basic Medical and Health Sciences, King's College London, and author of numerous books on ecology

Stephen Mustow is an environmental scientist at Ove Arup and Partners

Dr Ralph Oxley is a lecturer at the Centre for Arid Zone Studies, University of Wales, Bangor

Dr John Pernetta is the Project Manager of Land-Ocean Interactions in the Coastal Zone, a Core Project of the Netherlands Institute for Sea Research

Jonathan Spencer is a woodland ecologist working for English Nature

Nick Turland is an ecologist and botanist, working at the Natural History Museum, London